Current Topics in Environmental Health and Preventive Medicine

Series Editor
Hidekuni Inadera
Department of Public Health
University of Toyama
Toyama, Japan

Current Topics in Environmental Health and Preventive Medicine, published in partnership with the Japanese Society of Hygiene, is designed to deliver well written volumes authored by experts from around the globe, covering the prevention and environmental health related to medical, biological, molecular biological, genetic, physical, psychosocial, chemical, and other environmental factors. The series will be a valuable resource to both new and established researchers, as well as students who are seeking comprehensive information on environmental health and health promotion.

More information about this series at http://www.springer.com/series/13556

Tamie Nakajima • Keiko Nakamura
Keiko Nohara • Akihiko Kondoh
Editors

Overcoming Environmental Risks to Achieve Sustainable Development Goals

Lessons from the Japanese Experience

Editors
Tamie Nakajima
College of Life and Health Sciences
Chubu University
Kasugai, Aichi, Japan

Nagoya University
Nagoya, Japan

Keiko Nohara
Center for Health and Environmental Risk Research
National Institute for Environmental Studies
Tsukuba, Ibaraki, Japan

Keiko Nakamura
Department of Global Health Entrepreneurship
Tokyo Medical and Dental University
Tokyo, Japan

Akihiko Kondoh
Center for Environmental Remote Sensing
Chiba University
Chiba, Japan

ISSN 2364-8333 ISSN 2364-8341 (electronic)
Current Topics in Environmental Health and Preventive Medicine
ISBN 978-981-16-6251-5 ISBN 978-981-16-6249-2 (eBook)
https://doi.org/10.1007/978-981-16-6249-2

This Springer imprint is published by the registered company Springer Nature Singapore Pte Ltd.
The registered company address is: 152 Beach Road, #21-01/04 Gateway East, Singapore 189721, Singapore

Environmental Risk Subcommittee

Preface

Environmental risk is the impact or potential impact of pollution caused by an organization's activities on human health and the ecosystem. Originally, it covers all factors, including global environmental issues, but specifically, it often refers to the risks caused by hazardous chemicals.

Our country has a history of overcoming a half-century-long battle against pollution-related diseases, yet new environmental risks are also emerging. Since the end of 2019, Coronavirus disease (COVID-19) has been spreading around the world and is currently the biggest health concern. As of February 2021, more than 108.58 million people have been infected, and more than 2.39 million people died due to COVID-19. It is also worth noting the heavy use of plastic containers and medical supplies due to the pandemic. In this book, however, we concentrate on how Japan has overcome almost half a century of pollution to outline environmental risks we face today and discuss what actions we should take for the future, noting scientific and technological advances that have been made during this time.

Many chemicals have been initially developed to enrich our lives. However, their unregulated use, without the knowledge of their fate in living organisms and the ecosystem, has led to a variety of environmental and health hazards. In Japan, the term "pollution-related disease" is synonymous with disease. We have gained vast experience and knowledge since the recognition of four major pollution-related diseases (Minamata disease, Niigata-Minamata disease, Itai-itai disease, and Yokkaichi asthma) and arsenic poisoning-related disease in Toroku, Miyazaki. As a result, the concept of regulatory science, which involves assessing and managing the risks of hazardous substances in advance, and understanding these risks by the population, is important. Since then, many hazardous substances have been assessed and managed in regulatory science.

Nevertheless, challenges continue to arise, such as plastic marine pollution. Plastic is the most commonly used material for packaging and containers, and its use has alarmingly increased with the COVID-19 pandemic. Moreover, microplastics in the marine environment, along with polychlorinated biphenyls and other toxic chemicals, create an extremely complex problem. As a result, we have reached a point in time where we must question the costs of not only producing plastics but also using them. It is an urgent matter to review processes from "manufacture" to "use" and "disposal, treatment, and reuse," and minimize the emission of plastics into the environment. Above all, we must take advantage of the scientific and technological know-how of "Japan, the Land of Pollution."

In 2015, the United Nations adopted "Transforming our world: the 2030 agenda for sustainable development." It follows MDGs, which were set to be achieved by 2015. One of the characteristics of SDGs is that they target not only developing countries but also developed countries, and they aim to solve the world's sustainable development-related problems by 2030. This book is a summary of the activities of the Environmental Risk Subcommittee of the Science Council of Japan. Therefore, many important environmental issues may have been excluded. We hope that it will be possible to achieve SDGs by 2030 if all sectors and nations collaborate on a global scale.

Nagoya, Japan Tamie Nakajima
Kasugai, Japan

Contents

Chapter 1
Purpose of This e-Book

Tamie Nakajima

1.1 Introduction

In 2015, the United Nations declared "Transforming our world: the 2030 agenda for sustainable development," which aims to follow up millennium development goals (MDGs) and expand the efforts over the next 15 years targeting not only developing countries but also developed countries and aiming to solve the most pressing problems of the world regarding sustainable development by 2030.

To tackle nationwide problems, the Environmental Risk Subcommittee of the Science Council of Japan cosponsored two symposia: *What we think now, half a century after the certification of pollution-related diseases: achieving sustainable development goals (SDGs)* and another titled *Environmental circulation of hazardous substances and health: SDG 12—responsible consumption and production,* with the Japan Society for Hygiene and the Japan Society for Occupational Health, respectively, in 2019.

In this e-book, the history and current status of pollution-related diseases in Japan are described from the perspective of environmental risks, and the need for the development of new scientific technologies is highlighted. Next, we discuss how new technologies are being applied to environmental challenges in Japan. Consequently, we discuss the following seven SDGs: "good health and well-being" (Goal 3), "clean water and sanitation" (Goal 6), "sustainable cities and communities" (Goal 11), "responsible consumption and production" (Goal 12), "climate action" (Goal 13), "life below water" (Goal 14), and "life on land" (Goal 15). Besides, we discuss what it would take to achieve these Goals and reflect on the

T. Nakajima (✉)
Nagoya University, Nagoya, Japan

College of Life and Health Sciences, Chubu University, Kasugai, Aichi, Japan
e-mail: tnasu23@med.nagoya-u.ac.jp

T. Nakajima et al. (eds.), *Overcoming Environmental Risks to Achieve Sustainable Development Goals*, Current Topics in Environmental Health and Preventive Medicine, https://doi.org/10.1007/978-981-16-6249-2_1

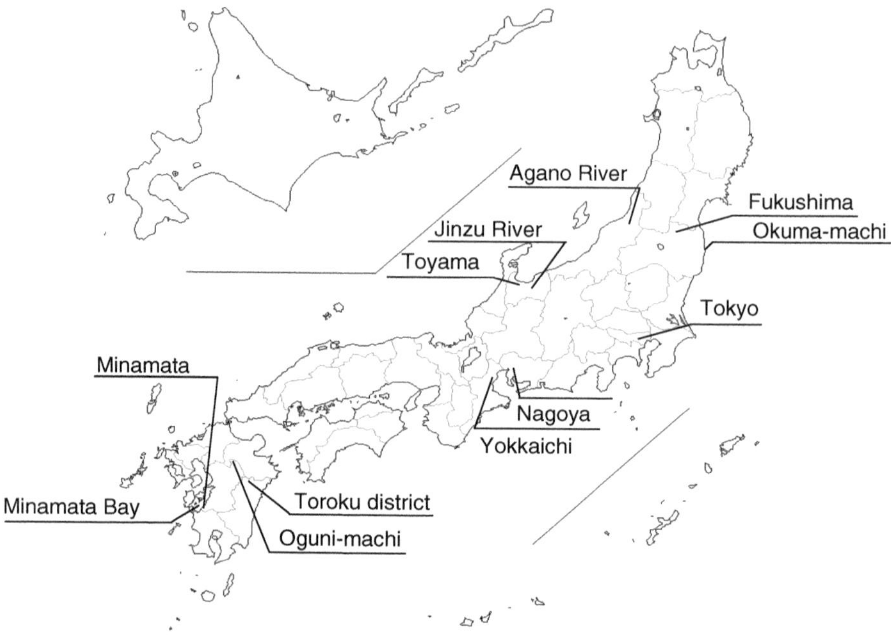

Fig. 1.1 The name of the places in Japan mentioned in each chapter of this e-book

creation of a sustainable and prosperous society. The names of the places mentioned in this e-book are shown in Fig. 1.1.

1.2 Part 1. Pollution-Related Diseases in Japan

Four major pollution-related diseases, Minamata disease, Niigata Minamata disease, *Itai-itai* disease, and Yokkaichi asthma, and arsenic pollution-related disease in Toroku were recognized in 1968 when Japan's gross domestic product was the second-largest in the world, and the number of complaints about pollution had increased rapidly from 20,000 to 80,000 a year. In this chapter, we provide an overview of pollution-related disease outbreaks (including their causes, environmental restoration, and the efforts of scientists and citizens) and the current situation in Japan.

These pollution diseases occurred amid the postwar reconstruction prior to the emergence of regulatory science (risk assessment, management, and communication). Although the cause of Minamata disease was methylmercury contained in factory effluent, it was not recognized as a pollution disease until 1968, twelve years after the first patient was diagnosed. Yokkaichi asthma, a respiratory disease caused by sulfurous gas emitted from a petroleum complex, was recognized as a public health hazard in 1965. The legal responsibility of each pollution-generating company was recognized and legal trials related to these pollution diseases were concluded. Afterwards, Minamata City and Yokkaichi City were designated as model cities by the government to accomplish new goals for city development and create basic principles by taking measures stated in their ordinances with participation of citizens, businesses, and the government that took half a century to achieve them.

Itai-itai disease was recognized as a pollution-related disease in 1968 despite having a history of more than 100 years. The disease was caused due to the pollution of the Jinzu River by cadmium discharged from the Kamioka Mining Works of Mitsui Kinzoku Mining Co. The full agreement between the company that caused *Itai-itai* disease and the patients went beyond national certification standards to provide a certain amount of relief for kidney damage caused by cadmium.[1] After a fight of more than 50 years for environmental rehabilitation by residents, the cadmium pollution burden of the Kamioka mine was mostly rectified by 2004.

Arsenic acid production began sometime between 1912 and 1926 (in the Taisho era) at a mine at Toroku, Miyazaki. In 1971, a teacher at an elementary school accused arsenic pollution-related disease in the Toroku district. Since then, the "Toroku Aersenic Pollution Disease" has been attracting attention. The river water flowing through this area was also contaminated with arsenic. In 1973, diseases caused by arsenic pollution were recognized. However, "lung cancer" caused by arsenic was discovered much later.[2] Even now, one or two people a year are diagnosed with cancer due to arsenic pollution, which differs from other pollution-related diseases. This is an overlap of pollution and industrial accidents. It has been difficult to restore the geological conditions quickly and fully eliminate the pollution, however.

These experiences have spawned regulatory science, which involves the study of the risks and management of hazardous substances. We have now entered the twenty-first century, the era of hazardous substance management.

1.3 Part 2. Environmental Risks of the Present

This part discusses how the theory of regulatory science (health risk assessment, risk management, and risk communication) is prevalent and understood based on four environmental risks facing our nation today. Polychlorinated biphenyls (PCBs), which are persistent organic pollutants (POPs), were the causative agents of the Kanemi rice oil disease (Yusho disease) that occurred over a wide area in western

[1] Nihon Keizai Shimbun: Overcoming national certification standards, compromise reached on settlement of Itai-itai disease, December 17, 2013 (in Japanese).

[2] Takahiko Kato, History of Kyushu Regional Meeting—Part 7—Chronic arsenic poisoning in Toroku, Takachiho-cho, Miyazaki Prefecture, Japan (2) Overlap of pollution and industrial accidents (occupational diseases) Japan Society for Occupational Health Kyushu Regional Meeting News No. 44, September 1, 2018 (in Japanese).

Japan in 1968. In 2001, the PCB Measures Law was enacted, and the chemical decomposition of PCBs started. However, the environmental and health risk assessment of newly designated POPs such as perfluorooctane sulfonic acid (PFOS) and perfluoronic acid (PFOA) has not yet been initiated.

The use of asbestos, which causes lung cancer and mesothelioma, was banned in 2012 in Japan. However, many people were exposed to this contaminant during the demolition of buildings. In December 2013, people were exposed to asbestos that was dispersed during its removal from a station building of the Nagoya Municipal Subway. A review committee conducted a health risk assessment and examined the safety of workers and residents. Thus, although the risk assessment of asbestos is now easily available, it is desirable to manage the risks of asbestos removal work.[3] Furthermore, the risks of asbestos exposure derived from serpentine rock and asbestos-scattering accidents persist, and in fact, mesothelioma cases have been reported.

Marine pollution by plastic waste is growing worldwide. Microplastics act as carriers of hydrophobic toxic chemicals, such as PCBs. Since the effects on the ecosystem and human health are not well-known, the government needs to urgently conduct basic research and epidemiological studies, and at the same time, develop technologies to recycle and dispose of the plastics that exist today. How we manufacture, use, dispose of, and recycle plastics must be urgently addressed.

The explosion at the Tokyo Electric Power Company's (TEPCO) Fukushima Daiichi Nuclear Power Station (NPP), which occurred following the Great East Japan earthquake, spread radioactive materials over a wide area in 2011 and adversely affected the surrounding environment and lives of residents. Residents who had evacuated the area after the accident have recently begun to return to their homes since the radiation levels have been deemed as "acceptable." It was interpreted that residents were psychologically trying to "understand," but not "consensus building." Ten years have already passed since the accident. It may be important that "understanding" requires not only scientific or economic rationality, but also empathy for the region and a shared philosophy of how society should be in the risk communication.[4]

---

[3] Study report on health and other measures related to asbestos scattering at Rokubancho Station, Nagoya City Transportation Bureau website, 2017 (in Japanese) https://www.kotsu.city.nagoya.jp/jp/pc/ABOUT/TRP0002011.ht

[4] Akihiko Kondoh. Solution and agreement in the Nuclear disaster: toward an inclusive society respecting relationship from sacrificial system. Trends in the Science 24(10); 49–52, 2019. (in Japanese).

1.4 Part 3. Toward a Sustainable Society

Measures must be taken to address environmental challenges and eliminate pollution-related diseases. An interesting case is the Minamata disease trial. The rationale behind the Supreme Court's decision to impose liability on not only the pollution-causing companies but also the national government and Kumamoto Prefecture reflects the concept of regulatory science. It shows that "the appropriate exercise of regulatory authority by the national and local governments over substances causing environmental risk problems" is necessary to achieve SDGs. For instance, in the past, priority was placed on plastic production, and no attention was paid to its disposal, which led to problems, such as dioxin generation. The current marine plastic pollution is also linked to microplastic pollution, and we hope that the emissions of these substances can be minimized by SDG efforts. In addition, as a "sustainable small local government model of harmony between the environment and the economy," Minamata City and Yokkaichi City have declared that they will overcome pollution and prioritize welfare. If done, this will be a model for the world as our country is striving to build a sustainable environment from the pollution of the past. Companies, meanwhile, are assessing the environmental impact of cleaning and developing advanced cleaning technologies, and Oguni Town, which was selected as a future city, is being built as a sustainable low-carbon city that makes use of local resources (geothermal energy and forests). An embryonic movement toward collaboration in various fields toward realizing SDGs is now being seen.

Part I
Learning from Case Analysis of Past Environmental Risks

Chapter 2
Minamata Disease

Katsuyuki Murata and Kanae Karita

Abstract Minamata disease, i.e., methylmercury poisoning, happened in the vicinity of Minamata Bay, Japan, by inhabitants' eating a large quantity of fish contaminated with wastewater discharged from a chemical plant. It was an environmental epidemic in a way but also a man-made calamity that occurred under the government's policy to put the highest priority on economic growth in the postwar era. Based on such a tragedy in Japan, an international treaty, i.e., Minamata Convention on Mercury, was designed to protect human health and the environment from anthropogenic emissions and releases of mercury and mercury compounds. This is exactly one effective way to achieve sustainable development goals (SDGs) for water and sanitation. By understanding the process of unraveling what the causal agent of Minamata disease was, people should recognize a mission to stay tuned toward the latest information on the globe, keep on watching the changing environment, and find out the truth without any prejudice.

Keywords Minamata disease · Didactic experience · Methylmercury poisoning Minamata Convention on Mercury · Fish-consuming populations · Sustainable development goals (SDGs) · Aquatic pollution

---

K. Murata (✉)
Department of Environmental Health Sciences, Akita University School of Medicine, Akita, Japan
e-mail: winestem@med.akita-u.ac.jp

K. Karita
Department of Hygiene and Public Health, Kyorin Unversity School of Medicine, Tokyo, Japan
e-mail: kanae@ks.kyorin-u.ac.jp

T. Nakajima et al. (eds.), *Overcoming Environmental Risks to Achieve Sustainable Development Goals*, Current Topics in Environmental Health and Preventive Medicine, https://doi.org/10.1007/978-981-16-6249-2_2

2.1 Introduction

Minamata disease, i.e., methylmercury poisoning, occurred in the vicinity of Minamata Bay, Kumamoto, Japan, by inhabitants' eating a large quantity of fish contaminated with wastewater discharged from a chemical plant. Methylmercury can easily enter the brain; therefore, adult-type Minamata disease patients suffered from damages to neurons in specific regions of the brain, mainly in the cerebrum and cerebellum, and they exhibited severe neurological symptoms. Besides, methylmercury taken into the body of a mother during pregnancy transferred to the fetus via placenta and caused damages to the fetal brain. Majority of fetal-type Minamata disease patients were born in 1955–1959 when the pollution was most severe. The babies had no abnormality during pregnancy or at delivery, but they showed symptoms similar to cerebral palsy such as disabilities of speech, head control, and walking. This chapter describes Minamata disease, a didactic experience in considering Sustainable Development Goals (SDGs) for "water and sanitation."

2.2 Health Effects of High-Level Methylmercury Exposure

Nippon Chisso Fertilizer Co., Ltd. (later renamed as Shin Nihon Chisso Fertilizer, hereinafter abbreviated as Chisso plant) started producing acetaldehyde in Minamata in 1932. Acetaldehyde production declined during World War II but increased again after the war [1]. From around 1953, strange events began to occur in the calm fishing village; for instance, cats went mad, ran around and jumped into the sea, and crows and birds living along the seashore fell from the sky and died [2–4]. Later on, some residents who had not experienced any health problems began to complain of a progressive numbness of the distal parts of the extremities and also of the lips and tongue. They also exhibited clumsiness of hands, speech disorder, hearing loss, and constriction of the visual field, in addition to gait disturbance that led to stumbling and staggering. Limitation in voluntary movements, often with convulsion and rigidity, was observed in severe cases. Some patients fell into a condition of clouding of consciousness or coma. Others suffered from generalized convulsions or were screaming day and night. In the worst cases, patients passed away in a month after showing the initial onset of symptoms.

In late April 1956, a 5-year-old girl with peculiar nervous symptoms from a fishing village on Minamata Bay visited the Chisso-affiliated hospital [2, 3]. Ten days later, her sister, aged 2 years, was also admitted to the hospital with clinical features similar to her elder sister. One pediatrician suspected that the children had an infectious disease, and consulted with Dr. Hosokawa, the Director of the hospital, saying "a patient with that strange symptom came again." Dr. Hosokawa had already examined two other infants with clinical manifestations similar to the two girls, but both had already died a couple of months after admission, without identification of the cause. The outbreak of these series of cases was reported by Dr. Hosokawa to the

Minamata Public Health Center on May 1, 1956, the day of the first official record of Minamata disease.

On May 28, 1956, a countermeasure committee for Minamata strange disease was organized [2–4]. The common features of cases with the strange disease were summarized as follows: The patients resided mainly in the fishing villages near Minamata Bay; most of them were fishermen or family members; they had eaten a large quantity of seafood even though they were neither fishermen nor their family; and, most of the patients were too poor to consult a doctor, and their meals depended on seafood but not rice, while most fishermen obtained rice by barter even under the monetary system. Three hypotheses on the cause were raised: (1) Japanese encephalitis; (2) poisoning resulting from well water; (3) food poisoning due to seafood that the cases ate. On August 24, 1956, the research group of Kumamoto University (a national university located close to Minamata) on Minamata disease was organized. In the first meeting on November 3, 1956, it was concluded that there were no clinical findings of inflammation and the bacteriological and virological tests were negative and that the disease was rather considered to be a poisoning due to heavy metals judging from symptoms of the patients. Table 2.1 summarizes the process leading

Table 2.1 History of Minamata disease

Year	Month	Events
1956	May	The patients with neurological symptoms of unknown cause were reported at the first time (official report of Minamata disease).
	August	The Minamata disease research group of Kumamoto University School of Medicine started field survey.
1957	February	Kumamoto University research group reported to Kumamoto prefecture "poisoning by seafood contaminated with a certain heavy metal is suspected."
	April	The director of Minamata Public Health Center fed cats with seafood from Minamata Bay and confirmed the onset of the disease.
	October	Health Science Research group sponsored by the Ministry of Health and Welfare announced the possibility of intoxication by selenium, manganese, or thallium.
1958	September	Kumamoto University research group considered organic mercury as a cause for the first time.
	September	Chisso plant changed drainage outlet from Minamata Bay to Minamata River.
1959	March	Patient onset near the mouth of the Minamata River
	October	Cat experiment was conducted in a hospital of Chisso plant, and acetaldehyde waste liquid caused the onset of the disease to the cat.
	November	The Food and Sanitation Investigation Committee reported to the Health and Welfare Minister that the cause was a certain organic mercury compound.
1960	April	A famed professor of the Tokyo Institute of Technology presented his hypothesis that noxious amine was to blame.
1963	February	Kumamoto University extracted methylmercury directly from acetaldehyde residue and announced that the causative substance was methylmercury.
1965	May	Official confirmation of an outbreak of Niigata Minamata disease.
1968	May	Chisso plant discontinued acetaldehyde production.
	September	The government officially announced that methylmercury compound derived from the Chisso plant was the causative substance of Minamata disease.

up to the government view of Minamata disease. Concerning details of the process of elucidating the causative substance, refer to the references [2–4].

2.3 Company's Responses

The wastewater from the Chisso plant which emptied into Minamata Bay had been suspected as the major source of a toxic substance contaminating fish and shellfish [3, 4]. Subsequent work had provided evidence that the responsible toxin was associated with the discharge of mercury-containing wastewater from the chemical plant. Dr. Hosokawa also fed food, to which wastewater from the plant had been added, to cats in July 1959, and his cat exhibited symptoms of Minamata disease. After his report, the plant ordered him to stop the research, and revealed his experimental result to the Ministry of International Trade and Industry and the Japan Chemical Industry Association, but not to the research group of Kumamoto University.

The Director of the Japan Chemical Industry Association visited Minamata and suggested that the explosive compounds from the former navy ammunition storage house were responsible for Minamata disease in September 1959, although this hypothesis had already been disproved in 1957 [3, 4]. Also, a famed professor of the Tokyo Institute of Technology presented his hypothesis that noxious amine was to blame in April 1960. Thus, the press had been placated with clever words by the Chisso plant side, and inane counterarguments by authorities advocating the company threatened the health and safety of the Minamata residents.

The formation of organic mercury from inorganic mercury used as a catalyst was reported in 1921; i.e., the mercury did not remain long in the form of a sulfate but was converted to an organic compound, and this acted as the catalyst [3, 5, 6]. In addition, there were two studies reporting several cases with organic mercury intoxication among workers engaged in acetaldehyde production from acetylene in 1930 and 1937 [7]. At the time of the outbreak of Minamata disease, these reports should have been available, and a few technicians of the chemical plant might have known about their findings because of their general knowledge of the catalytic transformation of acetylene. Nevertheless, the plant consistently insisted that they could not have foreseen the formation of organic mercury from inorganic mercury.

2.4 Government Response to Minamata Disease

After Minamata disease was suspected of being caused by eating fish and shellfish, the Kumamoto prefectural government began to discuss a ban on fish capture in Minamata Bay and also its consumption [2, 3]. The prefectural government decided to apply the Food Sanitation Act and referred to the Ministry of Health and Welfare for the propriety of application of the law on August 16, 1957. The reply from the

Ministry on September 11, 1957 was: "The inhabitants must be recommended to avoid eating the seafood caught in Minamata Bay because it may induce the strange disease affecting the central nervous function. However, there is no distinct evidence endorsing the toxicity of all fish and shellfish dwelling in the bay. Therefore, it appears to be impossible to apply the Food Sanitation Act to all the fish and shellfish caught in the relevant area."

Rapid economic growth in Japan since 1955 resulted in a high level of development of heavy chemical industries, great expansion in productivity, and rise in income levels. According to Prof. Reischauer, ex. US Ambassador to Japan from 1961 to 1966, industrial evolution in Japan was led by the Ministry of International Trade and Industry, giving the highest priority on Japan's industrial growth with strong authority in the government [8]. With that, the Ministry prioritized economic growth over certifying pollution or relieving patients from the outbreak of Minamata disease.

The Ministry of International Trade and Industry guided the Chisso plant on October 21, 1959 to stop draining water into the mouth of Minamata River but into Hyakken drain as before and to set up facilities for wastewater management by the end of the year [4]. The Ministry also instructed the government to inspect wastewater from all the plants producing acetaldehyde or vinyl chloride in Japan. However, results were not made public for fear that Minamata disease becomes a serious political issue [2]. As a consequence, the second disaster (i.e., Niigata Minamata disease) occurred in the Agano River basin; in July 1965, it was confirmed that there were 26 patients there and that five of them had died. The Government of Japan made a proclamation of its collective view regarding Minamata disease in the vicinity of Minamata Bay in Kumamoto prefecture and the Agano River basin in Niigata prefecture on September 26, 1968: "The methylmercury compound, secondarily generated in the process of production of acetaldehyde, was the causal agent of the disease" (Fig. 2.1). The operation of acetaldehyde production using the catalyst at the Chisso plant had ceased in May of that year. Twelve years had passed after the official report on Minamata disease. Control over industrial effluent based on the Factory Wastes Law was initiated in February 1969.

2.5 Issues Related to Officially Certified Patients of Minamata Disease

The reason for prolonged lawsuits for officially certified patients of Minamata disease was attributable to too strict diagnostic criteria, which was the same in the Morinaga dry-milk arsenic poisoning incident that occurred in that same period [9]. The criteria required specific clinical manifestations of Minamata disease or trials (i.e., pigmentation in skin, hepatomegaly, and anemia) in the Morinaga case. In addition, the criteria did not necessarily assume specific and objective diagnostic methods. The government was not aware of the causal agent of Minamata disease till 1959 and exposure assessment using biological samples of the patients did not take place. Before 1971 when cold vapor atomic absorption spectrometry method

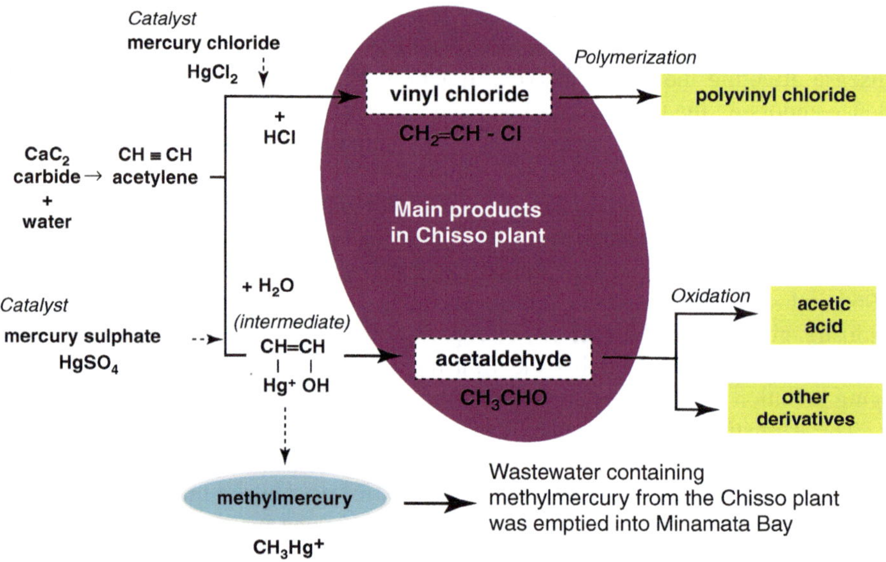

Fig. 2.1 Generation of methylmercury in the process of acetaldehyde production

was developed, Japanese researchers applied colorimetric methods with wet digestion and extraction with dithizone to analyze mercury in biological samples; also, they did not have any means for evaluating the sensitivity or precision of their analysis results. The fact that academic consensus on health effect, exposure assessment, and risk assessment of methylmercury still remained to be established, further exacerbated the tragedy of Minamata disease.

2.6 Health Effects of Low-Level Methylmercury Exposure

A discovery of fetal-type Minamata disease provided important evidence that intrauterine exposure to methylmercury affects the fetuses being the next generation [10]. A declined male birth ratio was observed among Minamata disease patients throughout Minamata city from 1955 to 1959 [10, 11]. Shared hypothesis was that a high level of methylmercury exposure during pregnancy caused excess miscarriage or stillbirth of male fetuses especially. This fact indicated that male fetuses were more susceptible to methylmercury than female ones [10–12]. Furthermore, it turned out that mercury concentrations in umbilical cord blood were about twice as high as those in maternal blood [10, 13], which explains the fact that mothers whose babies had fetal-type Minamata disease showed just a little symptoms of methylmercury poisoning. As a consequence, it was considered that fetuses were the most susceptible to environmentally hazardous substances, and Faroese Birth Cohort Study, Seychelles Child Development Study, and Japan Environment and Children's Study focused on the effect of prenatal exposures to chemical substances on child development [14].

In August 2005, the Food Safety Commission of the Cabinet Office in Japan reviewed the cohort studies results on health effects of methylmercury and identified that fetuses were the population with the highest risk to exposure of methylmercury contained in seafood [14]. The Commission replied to the Minister of Health, Labour and Welfare that the tolerable weekly intake (TWI) of methylmercury, without risk of adverse health effects, was 2.0 μg/kg body weight (BW)/week. On the basis of this TWI, "an advice about feeding fish and shellfish to pregnant women (i.e., fetuses)" was announced by one section of Food Sanitation Council established under the Pharmaceutical Affairs and Food Sanitation Council of Ministry of Health, Labour, and Welfare in November 2005.

An intervention study was conducted to scrutinize whether the provisional tolerable weekly intake (PTWI, 3.4 μg/kg body weight (BW)/week) of methylmercury in Japan was safe for adults [15, 16]. The experimental group was exposed to methylmercury at the PTWI level through consumption of bigeye tuna or swordfish for 14 weeks; for instance, a subject whose BW was 60 kg consumed approximately 200 g/week of these fish because the methylmercury concentration in fish was 1.0 μg/g. In the experimental group, mean hair mercury levels, determined before and after the dietary methylmercury exposure and after 15-week wash-out period following the cessation of exposure, were 2.30, 8.76, and 4.90 μg/g, respectively. The sympathovagal balance index of heart rate variability (HRV) was significantly elevated after the exposure, but it reverted to the baseline after the wash-out period. On the contrary, such changes in HRV parameters were not found in the control group. These results imply that long-term exposure to methylmercury in fish-consuming populations may become a potential risk for cardiac events originating from sympathovagal imbalance.

In Japan, tuna cut into slices (approximately 15 g per slice) is recommended as baby food by some websites because it has less small bones and is rich in iron. If a mother feeds 4–5 slices of tuna per week to her baby weighing 8 kg, the baby takes 6–9 μg/kg BW/week of methylmercury; that is, its intake would exceed the TWI mentioned above [14]. Infants and toddlers are not included in the dietary advisory of the Ministry of Health, Labour, and Welfare, whereas some countries warn all citizens not to eat certain fish (e.g., swordfish, spearfish, or shark). However, this does not mean that methylmercury intake by infants or toddlers does not affect the nervous system; merely, there exist no scientific reports having examined such study subjects. Further study is needed to clarify whether low-level methylmercury exposure not only *in uteri* but also in infancy causes irreversible damage to the developing central nervous system.

2.7 Mercury Pollution in the World

Aquatic pollution due to mercury/methylmercury has been repeatedly reported [17, 18]. Especially in gold mining in developing countries, gold ore is crushed and metallic mercury is added to make mercury-gold amalgam. Mercury vapor is then released into the air during the process of obtaining gold by directly heating the amalgam. Also, in some areas of mercury mines in China, mercury vapor is generated during the cinnabar refining [19]. Figure 2.2 presents workplaces that generate mercury vapor. Mercury vapor pollutes not only air but also soil and rivers nearby, resulting in mercury found in paddy rice grew downstream of the mines. Total mercury concentration was the highest in rice root, followed by stem, leaf, hull, and brown rice [17, 20]. Methylmercury was accumulated the most in root, followed by brown rice. On the basis of this finding, rice grown in contaminated fields, in addition to seafood, attracts attention as a special exposure source of methylmercury to humans. The mercury-gold amalgam method used as the gold extraction method pollutes air, soil, and rivers as described above [18]. A cyanidation method is an alternative, however, a large amount of cyanide released into rivers will affect biodiversity, reducing aquatic small animals. Recently, methods that do not use these harmful substances are under research, but it will take more time before they could be applied in developing countries. In some developing countries in South America and Africa, artisanal and small-scale gold miners

Fig. 2.2 Workplaces that generate mercury vapor

continue to use mercury [21], though medium- to large-scale miners may use such alternative methods; that is, communities and ecosystems around artisanal and small-scale gold miners, who are often not incorporated into formally legal or institutional structures, are continuing to be at risk. Mercury emissions associated with artisanal and small-scale gold mining account for almost 38% of the global total and are the major contributor to the emissions worldwide [21, 22]. Accordingly, transition away from mercury use is critical, while a reduction in use, safety measures, and management are encouraged.

Methylmercury in the aquatic environment is bioaccumulated through the food chains of plankton, shrimp, small to large fish, and marine mammals (such as whales) [23, 24]. Methylmercury exposure to humans occurs mainly by eating fish or shellfish. For that reason, inorganic mercury and mercury vapor contaminations in the aquatic environment can finally increase methylmercury exposure levels to humans. The Minamata Convention on Mercury is an international treaty designed to protect human health and the environment from man-made emissions and releases of mercury and mercury compounds. It was adopted and signed in Kumamoto on October 10, 2013 [25], and the treaty entered into force on August 16, 2017 (the United Nations Environment Programme reported that the signatory and ratification nations on March 31, 2021 were 128 and 131, respectively). The treaty aims to reduce the supply, use, emissions, and disposal of mercury with cooperation among developed and developing countries. The comprehensive measures at each stage of the above can reduce anthropogenic emissions of mercury and prevent global mercury pollution, including cross-border pollution. This is exactly one effective way to achieve SDGs. Especially, goals 3 and 14 would be involved in the above treaty. Goal 3 is to ensure healthy lives and promote well-being for all at all ages, implying that reducing the use of mercury-containing products and minimizing its releases will ultimately result in protecting human and environmental health. Next, Goal 14 is to conserve and sustainably use the oceans, seas, and marine resources for sustainable development; large coastal populations in every region depend on marine resources for their livelihoods, which are being threatened by the deterioration of coastal waters due to pollution. Many countries around the world should decrease the use and release of mercury from various land-based activities, prevent mercury from entering water sources, and reduce the accumulation of mercury in the food chain.

2.8 Conclusion

Minamata disease was an environmental epidemic in a way but also a man-made calamity that occurred under the government's policy to put the highest priority on economic growth in the postwar era. Today, high-level methylmercury pollution is not expected to occur in Japan. Still, a small amount of mercury is emitted from natural and anthropogenic activities (coal-fired power plants, etc.), and the westerlies carry it with yellow sand, implying that mercury could be transferred across the border from the Asian continent. Therefore, scientists must have a mission to stay tuned toward the latest information on the globe, keep on watching the changing environment, and find out the truth without any prejudice.

References

1. Arima S, editor. Minamata disease – studies during these 20 years and problems remaining today. Tokyo: Seirin-sha; 1979. (in Japanese)
2. Doi R. Minamata disease. In: Satoh H, editor. Toxicology today – from toxicology to science of biological control. Kyoto: Kinpodo; 1994. p. 93–108. (in Japanese).
3. Murata K, Sakamoto M. Minamata disease. In: Nriagu JO, editor. Encyclopedia of environmental health, vol. 3. Amsterdam: Elsevier; 2011. p. 774–80.
4. Social Scientific Study Group on Minamata Disease. In the hope of avoiding repetition of a tragedy of Minamata disease: What we have learned from the experience. Kumamoto: National Institute for Minamata Disease; 2001.
5. Vogt RR, Nieuwland JA. The role of mercury salts in the catalytic transformation of acetylene into acetaldehyde, and a new commercial process for the manufacture of paraldehyde. J Am Chem Soc. 1921;43(9):2071–81.
6. Ishihara N. Bibliographical study of Minamata disease. Nihon Eiseigaku Zassshi. 2002;56(4):649–54. (in Japanese with English abstract)
7. Koelsch F. Gesundheitsschadigungen durch organische Quecksilberverbingdungen. Arch Gewerbepathol Gewerbehyg. 1937;8:113–6.
8. Reischauer EO. The Japanese. Tokyo: Charles E Tuttle Co; 1977.
9. Dakeishi M, Murata K, Grandjean P. Long-term consequences of arsenic poisoning during infancy due to contaminated milk powder. Environ Health. 2006;5:31.
10. Sakamoto M, Nakamura M, Murata K. Mercury as a global pollutant and mercury exposure assessment and health effects. Nihon Eiseigaku Zasshi. 2018;73(3):258–64. (in Japanese with English abstract)
11. Sakamoto M, Nakano A, Akagi H. Declining Minamata male birth ratio associated with increased male fetal death due to heavy methylmercury pollution. Environ Res. 2001;87(2):92–8.
12. Tatsuta N, Murata K, Iwai-Shimada M, Yaginuma-Sakurai K, Satoh H, Nakai K. Psychomotor ability in children prenatally exposed to methylmercury: the 18-month follow-up of Tohoku study of child development. Tohoku J Exp Med. 2017;242(1):1–8.
13. Sakamoto M, Murata K, Kubota M, Nakai K, Satoh H. Mercury and heavy metal profiles of maternal and umbilical cord RBCs in Japanese population. Ecotoxicol Environ Saf. 2010;73(1):1–6.
14. Murata K, Iwata T, Maeda E, Karita K. Dilemma of environmental Health Research. Nihon Eiseigaku Zasshi. 2018;73(2):148–55. (in Japanese with English abstract)
15. Yaginuma-Sakurai K, Murata K, Shimada M, Nakai K, Kurokawa N, Kameo S, Satoh H. Intervention study on cardiac autonomic nervous effects of methylmercury from seafood. Neurotoxicol Teratol. 2010;32(2):240–5.
16. Karita K, Iwata T, Maeda E, Sakamoto M, Murata K. Assessment of cardiac autonomic function in relation to methylmercury neurotoxicity. Toxics. 2018;6(3):38.
17. Murata K, Yoshida M, Sakamoto M, Iwai-Shimada M, Yaginuma-Sakurai K, Tatsuta N, Iwata T, Karita K, Nakai K. Recent evidence from epidemiological studies on methylmercury toxicity. Nihon Eiseigaku Zasshi. 2011;66(4):682–95. (in Japanese with English abstract)
18. Karita K, Sakamoto M, Yoshida M, Tatsuta N, Nakai K, Iwai-Shimada M, Iwata T, Maeda E, Yaginuma-Sakurai K, Satoh H, Murata K. Recent epidemiological studies on methylmercury, mercury and selenium. Nihon Eiseigaku Zasshi. 2016;71(3):236–51. (in Japanese with English abstract)

19. Iwata T, Sakamoto M, Feng X, Yoshida M, Liu X-J, Dakeishi M, Li P, Qiu G, Jiang H, Nakamura M, Murata K. Effects of mercury vapor exposure on neuromotor function in Chinese miners and smelters. Int Arch Occup Environ Health. 2007;80(5):381–7.
20. Meng M, Li B, Shao JJ, Wang T, He B, Shi JB, Ye ZH, Jiang GB. Accumulation of total mercury and methylmercury in rice plants collected from different mining areas in China. Environ Pollut. 2014;184:179–86.
21. United Nations Environment Programme. Analysis of formalization approaches in the artisanal and small-scale gold mining sector based on experiences in Ecuador. Mongolia, Peru, Tanzania and Uganda: Tanzania Case Study; 2021. https://wedocs.unep.org/handle/20.500.11822/31275
22. United Nations (UN) Environment. Artisanal and small-scale gold mining. Chap 3. In: Global emissions of mercury to the atmosphere from anthropogenic sources. Global mercury assessment technical background report 2018. Geneva: UN Environment Programme, Chemicals and Health Branch; 2019. https://www.unep.org/resources/publication/global-mercury-assessment-2018.
23. National Research Council. Toxicological effects of methylmercury. Washington, DC: National Academy Press; 2000.
24. Grandjean P, Satoh H, Murata K, Eto K. Adverse effects of methylmercury: environmental health research implications. Environ Health Perspect. 2010;118(8):1137–45.
25. Sheehan MC, Burke TA, Navas-Acien A, Breysse PN, McGready J, Fox MA. Global methylmercury exposure from seafood consumption and risk of developmental neurotoxicity: a systematic review. Bull World Health Organ. 2014;92(4):254–69.

Chapter 3
Itai-Itai Disease: Health Issues Caused by Environmental Exposure to Cadmium and Residents' Fight to Rebuild the Environment in the Jinzu River Basin in Toyama

Keiko Aoshima

Abstract In the Jinzu River Basin of Toyama Prefecture, the operation of a large-scale lead-zinc mine and smelting plant (Kamioka Mine) located upstream, from the 1890s to 1970s, released large volumes of slag and wastewater, which polluted the environment with heavy metals, such as cadmium. Rice paddies with an area of 1500 ha were contaminated with heavy metals as those soils used water from the Jinzu River for irrigation. The polluted irrigation water was also used for drinking; hence, residents of the Jinzu River Basin developed a chronic cadmium poisoning symptom, locally known as the *itai-itai* disease. In this chapter, we discuss the history of the mining pollution in the Jinzu River Basin that lasted for over 100 years, proceedings for the damages by the *itai-itai* disease intended to clarify the liability of the mining company, and history and current situation of remediation of the environment contaminated with cadmium through inspection of the mine by local residents, which was granted to them as the result of winning the case.

Keywords Cadmium · Countermeasures · Environmental pollution from the Kamioka Mine · *Itai-itai* disease · Tackling by local residents

---

K. Aoshima (✉)
Hagino Hospital, Toyama, Japan
e-mail: keiaoshima@po4.nsk.ne.jp

T. Nakajima et al. (eds.), *Overcoming Environmental Risks to Achieve Sustainable Development Goals*, Current Topics in Environmental Health and Preventive Medicine, https://doi.org/10.1007/978-981-16-6249-2_3

3.1 Introduction

"Since the beginning of time, rivers are one of the blessings bestowed by nature that are essential not only for transportation and irrigation but also drinking and many other aspects of life, which we took for granted, and the Jinzu River is no exception. In recent years, rivers have become polluted with waste inevitably generated via companies' economic activities, causing the destruction of their natural environment. In the absence of established regimes for maintaining and protecting the natural environment, such as rivers, building a facility equipped with a technology that can prevent the damages from the said wastes and relief from the damages due to the adequacy thereof must be demanded from the company that engages in the economic activities in question" (excerpt from the judgment of first instance of *itai-itai* disease trial, 1971) [1].

Itai-itai disease (meaning "it hurts, it hurts" disease), which is caused by cadmium pollution, is the most severe type of chronic cadmium poisoning, and it has a condition of proximal renal tubular osteomalacia; it is prevalent in the Jinzu River Basin of Toyama Prefecture [2]. The environmental contamination with cadmium was a result of the operation of the Mitsui Kamioka Mining and Smelting Co., Ltd. (hereinafter, Mitsui), which was located along the Takahara River in the upper Jinzu River [3, 4]. However, since the company had not accepted any responsibility for a long time, patients suffered from *itai-itai* disease and their families sued the company to pursue its liability. In March 1968 patients and bereaved families filed a claim for damages against Mitsui in the Toyama District Court. The trial lasted for 3 years and 4 months, and the verdict of the first instance was handed down by Chief Justice Toshio Okamura on June 30, 1971. The document detailing the ruling that included the well-known portion mentioned at the beginning [1] was read. Although Mitsui appealed the verdict, the residents won the appeal on August 9, 1972, in the Nagoya High Court, Kanazawa Branch. Victimized residents and their legal team negotiated directly with Mitsui at their Tokyo headquarters on August 10, the day after the verdict, and signed "Pledge on the compensation for *itai-itai* disease," "Pledge concerning the soil pollution problems," and "Agreement on pollution control" regarding on-site inspections. Residents' activities based on the pledge and agreement became the starting point of environmental reclamation [3].

In this chapter, we discuss the history and current situation regarding the *itai-itai* disease and introduce the achievements of the environmental reclamation activities by the residents, especially from the viewpoint of overcoming water pollution, based on the report by Hata [5].

3.2 Environmental Pollution Due to Cadmium and Its Damages

From the 1890s to the 1970s, in the Jinzu River Basin of Toyama Prefecture, slag and wastewater discharged from a large-scale lead-zinc mine and smelting plant (Kamioka Mine) located upstream polluted the environment with heavy metals, such as cadmium. As the result of the long-term intake of cadmium through ingestion of agricultural crops, such as rice, and drinking of irrigation water, local residents developed chronic cadmium poisoning symptoms, most notably, the *itai-itai* disease [2, 6]. The Fig. 3.1 shows natural history from the accumulation of cadmium in the body tissues to the development of *itai-itai* disease. It was speculated that the symptoms of the *itai-itai* disease appeared in the area around the 1910s, and the disease became prevalent by 1935. However, it was the 1960s, almost 50 years after the first appearance in the area, when the media picked up the disease as a social problem and the government began investigating it [6–8].

3.2.1 Development of the Kamioka Mine and History of the Damage

After the Meiji Restoration in 1868, the Japanese government vigorously pursued the development of mining and manufacturing industries to catch up with the western developed nations and modernization. Mitsui, a leading

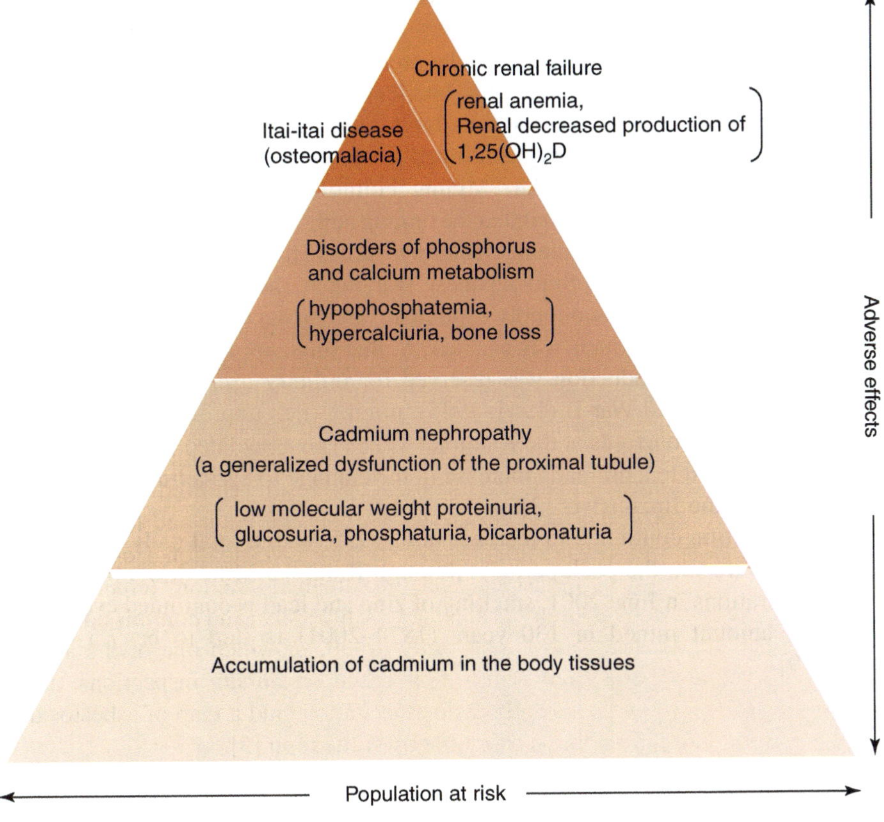

Fig. 3.1 Natural history of *itai-itai* disease (Aoshima, unpublished)

conglomerate in Japan, began full-scale operations and development of the Kamioka Mine in 1886. In the early stages of the development, the main product of mining and smelting was silver with lead as a by-product. In 1905, a year after the start of the Russo-Japanese War, they began mining zinc ore, which was previously discarded as a contaminant during lead smelting. Between the 1910s and 1920s, including World War I (1914–1918), the production of lead and zinc at Kamioka Mine rapidly increased up to 60% to 70% of the total national production. In 1909, they introduced a flotation method, in which ores were crushed and granulated, to separate lead and zinc ores. In 1927, Mitsui adopted an all-mud priority flotation method. As a result, large volumes of wastewater and slag containing granulated heavy metals were generated. Owing to the inadequate facility and process of waste disposal, the wastes were released in the Takahara River that ran immediately below the mine and smelting plant. In 1917, the residents of the downstream area had claimed damages to farm crops, fruit-bearing trees, and woodlands, and received compensation from Mitsui. In 1919, those residents took samples of the soil irrigated with the water from the Jinzu River and requested to the then Agricultural Experiment Station of the Ministry of Agriculture and Commerce to analyze them for lead, copper, and zinc. The results showed that the zinc concentration in the soil was particularly high. With this evidence, they demanded the mining authorities to take action [3].

After the Great Depression of 1929, the production of lead and zinc dramatically decreased; however, the Manchurian Incident in 1931 triggered the increase. Although the company for the first time built a tailings pond in 1931, the damage was not mitigated; rather, it only continued to increase. In 1932, affected farmers and fishery operators formed the "Mining Pollution Eradication Association in Jinzu River" and repeatedly demanded the mining authorities and Toyama Prefecture to conduct surveys and implement countermeasures [3].

In 1937, after the Second Sino-Japanese War began, the Japanese economic system transitioned to a state monopoly capitalism, and an act on increased production of important mining products was enacted. At this time, approximately 60% of the total lead and zinc production was reserved for military use. Between 1931 and 1944, including World War II (1941–1945), mining (i.e., unprocessed ore production) increased by fourfolds at the Kamioka Mine. The associated increase in waste and wastewater led to a dramatic increase in the damage to agriculture and fisheries downstream of the Jinzu River [3].

The production continued to increase until 1975, with a total daily mining volume of 6600 tons in the peak period. Although Kamioka Mine discontinued their mining operations in June 2001, smelting of zinc and lead is continued even today. The total amount mined in 130 years (1874–2001) is said to be 7.5 million tons [6, 9].

3.2.2 Health Impact on the Residents of the Jinzu River Basin: History of itai-itai Disease

The onset of the *itai-itai* disease is believed to be in the 1910s, and the disease became prevalent during the 1935–1944 period. However, under the social situation during the war where "mining pollution" was a taboo word, there was no administrative measure taken. In 1946, after World War II, the residents finally expressed that "a strange disease was rampant" and requested Kanazawa University to conduct a survey. However, it approximately took another 10 years until it was recognized as a social problem, and it was in 1955 when the name "*itai-itai* disease" was first made public [6–8].

Early researches focused on clinical and medical research including clinical findings, pathology, diagnosis, and treatment, and the causes were assumed to be advanced age, malnutrition, prolificness, and hard labor. Although some pointed out that mining pollution may be the cause, there was no specific survey or study to prove it. Under such circumstances, Dr. Kin-ichi Yoshioka, Ph.D. in agriculture and economics, conducted a survey in collaboration with Dr. Noboru Hagino, the director of Hagino Hospital (since the *itai-itai* disease was rampant in the surrounding area of the hospital at that time), and in June 1961, Dr. Yoshioka published "A Study Report on the Mining Pollution in the Jinzu River Basin: Mining-related Damages on Agriculture and People (*Itai-Itai* Disease) [10]," pointing out, for the first time, the possibility of cadmium poisoning [6–8].

Following this publication, the national government and government of Toyama Prefecture initiated investigations and studies in the 1962–1963 period. Based on the results of those investigations and studies, the national government (the Ministry of Health and Welfare) announced in May 1968 that "*itai-itai* disease is chronic poisoning by cadmium discharged from Kamioka Mine, and is a pollution-related illness." It was then when countermeasures by the national government and administrations finally began [6–8].

3.3 Recent Trends Regarding the *Itai-Itai* Disease

Itai-itai disease is the most severe type of chronic cadmium poisoning. In the early stages, proximal renal tubular dysfunction (cadmium nephropathy) is developed, and glucose, amino acids, phosphoric acid, uric acid, and low molecular weight proteins such as β_2-microglobulin are excreted into urine in increased amounts (Fig. 3.1). Cadmium accumulates in the proximal tubule and the liver for a long time, and even today, multiple proximal renal tubular dysfunction is rampant and persists among elderly residents, aged 80 years or above, who live in the cadmium-contaminated areas of the Jinzu River Basin regardless of their sex [4, 7]. A recent survey by the Ministry of the Environment reported that there were about 1000 cases of proximal renal tubular dysfunction with a urinary β_2-microglobulin (U-β_2-MG) of $\geq$1 mg/gCr [11]. However, the national government (Ministry of the Environment) acknowledges only the *itai-itai* disease as a pollution-related illness, and cadmium nephropathy is not recognized as such [6]. Under such circumstances, in December 2013, victimized residents forced Mitsui, through direct negotiation, to compensate those with cadmium nephropathy (U-β_2-MG of 5 mg/gCr or higher) with a lump sum of 600,000 yen [4, 7].

In our recent study [4], the patients who persistently suffered from hypophosphatemia among those who had the severe case of proximal renal tubular dysfunction (U-β_2-MG of 20 mg/gCr or higher) developed osteomalacia; i.e., the *itai-itai* disease. In addition, in patients who received an active vitamin D treatment, the onset of osteomalacia was suppressed, but the kidney function continued to deteriorate, which in turn led to the onset of chronic renal failure and renal anemia. A prognosis survey conducted for 10 years and 10 months from June 2009 to March 2020 showed that 12 out of 18 subjects with U-β_2-MG of 20 mg/gCr or higher (severe case) had died, and the cause of death of a half of them (six patients) was a chronic renal failure [12].

Since the start of the certification system for the *itai-itai* disease in 1967, the certified patients have amounted to 200 (195 women and 5 men) by January 2021. As of the end of March 2021, only one (female) was still alive. However, even after 2000, 29 subjects (26 women and 3 men) applied for the certification, of which 17 (15 women and 2 men) were certified. It is a pollution-related illness that continues even in the twenty-first century.

3.4 Fifty Years of Residents' Fight to Reclaim the Environment and Its Results

3.4.1 Field Survey of Kamioka Mine Based on the Pollution Prevention Agreement

The trial for the *itai-itai* disease was a proceeding for damages pursuant to the provision of no-fault compensation in the Mining Act, and the compensation for damages was paid out as a result of overall victory for the plaintiffs (patients and bereaved family). However, reclaiming the contaminated environment and taking countermeasures against the source of release remained to be major challenges [3, 5]. The Pollution Prevention Agreement signed a day after the verdict states that if the victimized residents deem it necessary, residents or designated experts can conduct an on-site investigation into relevant facilities in the Kamioka Mine, such as the final wastewater treatment facilities; including drainage ditches and the waste slag dumping ground, and voluntarily collect relevant information [3]. This on-site investigation based on the Pollution Prevention Agreement have been conducted every year since 1972, and as of 2020, more than 8500 residents have participated in 49 investigations. Residents have commissioned their own studies and investigations into the sources of release, such as wastewater and stack gas, made recommendations to the Kamioka Mine based on the results, and realized pollution prevention measures. Furthermore, in 2015, they established the Special Committee on Countermeasures against sources of release, consisting mainly of residents. This committee is analyzing the data in "annual reports" on the pollution prevention measures conducted by Kamioka Mine and regarding the on-site investigations. A case example that has dramatically enhanced pollution prevention measures based on the Pollution Prevention Agreement between an affected residents' group and a company that has released a pollutant is unprecedented not only in Japan but in the world [5].

3.4.2 Result of Measures Against the Sources at Kamioka Mine

The Kamioka Mine operates mining, ore dressing, and smelting, and all these processes require a large volume of water. The characteristics of water used for mining industry are that the source is decentralized (e.g., mountain streams in mountainous regions, infiltrated water, mine water; and river water to a certain extent), area where the water is used is expansive, and thereby, unlike factories located in flat areas, it is extremely difficult to employ intensive modern wastewater treatment methods. The first and second on-site investigations in 1972 and 1973 showed that approximately 35 kg/month of cadmium was released from eight drains from the Kamioka Mine into the Jinzu River, and approximately 5 kg/month of cadmium in stack gas was released into the atmosphere [3, 5, 13]. Thus, turbidity separation of the mine water, recycling of the ore dressing process water, improvements to the wastewater treatment facilities, and treatment of seepage from the waste slag dumping ground were performed as countermeasures against wastewater. As a result, the amount and concentration of cadmium released from eight (currently seven) drainages from the Kamioka Mine decreased from ca. 35 kg/month and 9 ppb in 1972 to ca. 3 kg/month and 1 ppb in 2015 [3, 5, 14]. As countermeasures against stack gas, not only the collection of dust from furnaces but also enhancement to the collection of environmental dust within buildings and improvements to the processes that had high cadmium emissions were implemented. As a result, the amount of cadmium released with the stack gas decreased from ca. 5 kg/month in 1972 to ca. 0.1 kg/month [3, 5, 13, 14].

In addition to the wastewater and stack gas, water and sediment in the Jinzu River Basin got contaminated due to the release of heavy metals from dormant or abandoned mines, spoil tips, and along old tram tracks [3, 13]. After understanding the situation through aerial photographs and field surveys, separation of water from contaminated streams and water from noncontaminated streams, covering with soil, tree planting, and collection and treatment of contaminated groundwater were performed as countermeasures against dormant or abandoned mines and spoil tips. In this manner, the amount of cadmium released from dormant or abandoned mine and spoil tips was reduced to approximately 1 kg/month [5, 14].

3.4.3 *Improvements to the Quality of Water and Sediment in the Jinzu River Basin*

The results of the investigations into the quality of water and sediment in the entire Jinzu River Basin showed that the cadmium concentrations in the water and sediments in upstream from the Kamioka Mine and noncontaminated tributaries were 0.1 ppb and 0.05 ppm, respectively, while those in downstream from the Kamioka Mine and contaminated tributaries were 0.2–10 ppb and 1–10 ppm, respectively, clearly indicating the pollution load of the Kamioka Mine on the Jinzu River [3, 5, 13]. In particular, the investigations into the amount of cadmium released in the Jinzu River Basin conducted between 1975 and 1977 showed that the majority of the pollution load from the Kamioka Mine was not from eight drains and dormant or abandoned mines but from contaminated groundwater gushing out into the canal for power generation installed for the Hokuriku Electric Power Company (hereinafter, Hokuriku Electric canal) that runs under the Mine [3, 5, 13]. As countermeasures against the pollution load of the Hokuriku Electric canal, a system was installed to collect sump water from the canal, and water was pumped up from barrier wells. As a result, the amount of cadmium released was decreased from ca. 21 kg/month in 1977 to ca. 0.5 kg/month in 2015 [3, 5, 14].

Since 1980, water quality in the Jinzu River Basin is continuously monitored. Water is automatically sampled and analyzed every 3 hours every day at four locations: Higashimachi Power Plant (later changed to Tono canal) at the entrance of the Hokuriku Electric canal, Maki power plant at the exit of the Hokuriku Electric canal, Jinzu River Dam No. 1 power plant, and Ushigakubi irrigation water. Ushigakubi irrigation water is especially important, and analysis results are cross-checked between those analyzed by the Kamioka Mine and those analyzed by the residents. In 1980, the cadmium concentrations at Maki power plant, Jinzu River Dam No. 1 power plant, and Ushigakubi irrigation water were 0.2 ppb or higher. Since 1996, those concentrations have been below 0.1 ppb–the level equivalent to that in noncontaminated river. Since 2004, the cadmium concentrations in Tono canal in the upstream of the Kamioka Mine and Ushigakubi irrigation water have been 0.07 ppb and deemed that the pollution load from the Kamioka Mine is not zero, but has become negligible. As such, the aim to restore the water quality in the Jinzu River to the background level is achieved [5].

The cadmium concentration in dam sediments in the downstream of the Jinzu River decreased from 2 ppm to 3 ppm in 1975 by 10 times to 0.2–0.3 ppm in 2015, showing that the soil quality improved to the background level (0.5 ppm or below) [5, 14].

3.5 Conclusions

A long-lasting project [33 years (1979–2012)] of reclamation of paddy soils contaminated with cadmium, which reclaimed 863 ha of paddy soils, was completed [7], although this chapter did not discuss this project further. Extensive and severe environmental pollution with heavy metals, damages to the agricultural and fishery industries caused by the development of the Kamioka Mine by Mitsui, and a tenacious fight of the residents in the Jinzu River Basin regarding the damages to humans, which lasted for over 100 years, especially the fight in the latter half that began with the trial, are noteworthy. There remains for us to learn about the painstaking works of the people who endeavored to reclaim the environment and their achievements.

Conflict of Interest The author has no conflict of interest to declare.

References

1. Toyama District Court Civil Division, the Presiding Judge Toshio Okamura, Judge Hideo Ohashi, and Judge Masayuki Sano. 1968 (Wa) No. 41. Verdict. Tokyo: General Library; 1973. p. 5–147. (in Japanese).
2. Saito H, Aoshima K. Heavy metals & human body II- the impact of heavy metals on human body. In: Chino M, Saito H, editors. Heavy metals & biology. Tokyo: Hakuyusha Publishers; 1988. p. 167–211. (in Japanese).
3. Kurachi M, Tonegawa H, Hata A, editors. Mitsui capitals and Itai-Itai disease. Kyoto: Otsuki Shoten; 1979. (in Japanese).
4. Aoshima K. Recent clinical and epidemiological studies of itai-itai disease (cadmium-induced renal tubular osteomalacia) and cadmium nephropathy in the Jinzu River basin in Toyama prefecture, Japan. In: Himeno S, Aoshima K, editors. Cadmium toxicity. Singapore: Springer; 2019. p. 23–37.
5. Hata A. 45 years of measures against the source. In: Editorial Committee of the 50th Anniversary of the Itai-itai Disease Countermeasures Council, editor. Itai-itai Disease: overcoming difficulties over 100 years. Toyama: Itai-itai Disease Countermeasures Council; 2016. p. 234–41. (in Japanese).
6. Matsunami J. The 100 years of damages caused by cadmium: retrospective and outlook. Toyama: Katsura Shobou; 2015. (in Japanese).
7. Aoshima K. Itai-itai disease: renal tubular osteomalacia induced by environmental exposure to cadmium- historical review and perspectives. Soil Sci Plant Nutr. 2016;62:319–26.
8. Aoshima K, Horiguchi H. Historical lessons on cadmium environmental pollution problems in Japan and current cadmium exposure situation. In: Himeno S, Aoshima K, editors. Cadmium toxicity. Singapore: Springer; 2019. p. 3–19.
9. Kamioka Mining and Smelting Co., Ltd. Bulletin "Kozan", 2003. 37-39. (in Japanese).
10. Yoshioka K. A study report on the mining pollution in the Jinzu River Basin: mineral-related damages on agriculture and people (itai-itai disease). 1961. (in Japanese). In: Yoshioka K, editor. Study of Itai-itai disease. Yonago: Tatara Shoboh Publisher; 1970. p. 5–95.
11. Ministry of the Environment. Report on cadmium contaminated area resident health impact survey committee. 2019. (in Japanese).
12. Aoshima K. A retrospective, observational study of cadmium nephropathy in the inhabitants of the Jinzu River basin, Toyama, Japan: mortality and causes of death. Jpn J Hyg. 2021;76:S169. (in Japanese).
13. Research Group Commissioned by the Representative Council for Cadmium Poisoning Victims in the Jinzu River Basin. Measures against sources of itai-itai disease at Kamioka Mine and Smelting after the court case: comprehensive report by the commissioned research group. 1978. (in Japanese).
14. Kamioka Mining and Smelting Co., Ltd. The 2015 mining-related pollution prevention measures by Kamioka Mining. 2016. (in Japanese).

Chapter 4
Arsenic Intoxication in Toroku District of Japan

Yoshiki Kuroda

Abstract As you can see from the large amount of arsenic contained in the hot springs and releasing arsenic from volcanoes, the crust and soil are often contaminated by arsenic. Arsenic takes two forms (organic and inorganic compounds) in the environment and inorganic arsenic is highly harmful. Since ground water is polluted by inorganic arsenic; it could be unsafe to human health. In developing countries, groundwater is frequently used as drinking water to prevent infectious diseases such as gastrointestinal infections. Therefore, many countries are suffering from arsenic contamination.

In Japan, many people use water from the river, but not groundwater, as drinking water. Therefore, the occurrence of diseases due to arsenic-contaminated groundwater is not common. However, historically, many patients suffered from arsenic poisoning around some mines. The most popular case was that of arsenic pollution in the Toroku mine, Miyazaki prefecture. I introduce the arsenic pollution concerning the Toroku mine.

Keywords Arsenic · Arsenous acid · Toroku · Matsuo · Miyazaki · Mine

4.1 Introduction

The atomic number of arsenic is 33 and it has three allotropes. In the environment, arsenic exists as a metal. However, it also exists as inorganic or organic arsenic compounds. The main inorganic arsenic compounds are trivalent hydrogen arsenide (arsine), arsenic trichloride, diarsenic trioxide (arsenous acid), arsenic pentoxide,

---

Y. Kuroda (✉)
Faculty of Medicine, University of Miyazaki, Miyazaki, Japan
e-mail: ykuroda@med.miyazaki-u.ac.jp

T. Nakajima et al. (eds.), *Overcoming Environmental Risks to Achieve Sustainable Development Goals*, Current Topics in Environmental Health and Preventive Medicine, https://doi.org/10.1007/978-981-16-6249-2_4

and hydrates of these compounds. Organic arsenic compounds are synthesized in living organisms. In addition, artificial arsenic compounds are used as pesticides and pigments. Arsenic absorbed in the body is methylated and becomes organic arsenic, which is rapidly excreted in the urine [1]. When the arsenic exposure level is low, the arsenic absorbed in the body is metabolized via this metabolic pathway, and side effects are minimal. However, if humans are exposed to high levels of arsenic, unmetabolized arsenic inhibits enzyme receptors with thiol groups [2]. In addition, oxidative stress superoxide is generated in the body, and unmetabolized arsenic consecutively induces genotoxic and non-genotoxic effect [3]. Acute arsenic poisoning is characterized by digestive system symptoms, neuropsychiatric symptoms, liver dysfunction, and cardiovascular disorders. Chronic exposure to arsenic could induce skin lesions and malignant diseases such as skin cancer, lung cancer, and bladder cancer.

4.2 Arsenic Pollution Worldwide

Many countries are suffering from arsenic contamination worldwide [4–8]. Developing countries such as Chile [9], Argentina [10], Bangladesh [11–13], and China [14, 15] are severely distressed by arsenic contamination. In Chile and Argentina, rivers are polluted with arsenic, and contaminated river water is often used as drinking water. As a result, serious arsenic poisoning issues have been reported in these countries. In contrast, since groundwater is safe from microbial contamination (such as pathogenic digestive tract bacteria), well water has been frequently used and is recommended in many counties where the water supply is inadequate. However, groundwater could be a risk factor for arsenic contamination and arsenic poisoning. In China, arsenic poisoning is caused by the indoor use of coal, which contains arsenic [16]. Many countries in Southeast Asia also suffer from arsenic pollution [17–21]. It is thought that the source of arsenic could be the Himalayas, and the downstream sediments contain high levels of arsenic, which contaminates pipe-borne water.

4.3 Arsenic Pollution in Japan

In Japan, surface water (river water) has been habitually used as drinking water, and groundwater has rarely been used for this purpose. Therefore, there are few cases of arsenic poisoning caused by drinking water. Since there are many volcanoes, Japan is known as a volcanic country. It is said that the arsenic concentration in marine products such as soft seaweed [22], hijiki [23–25], and kelp near Japan is high. In particular, hijiki contains high quantities of inorganic arsenic [22, 25]; therefore, it could be dangerous if eaten in large quantities. Since ancient times, a large amount of ore (arsenopyrite) containing high quantities of arsenic has been produced in

Fig. 4.1 Ruin of arsenic kiln (provided by Miyazaki Prefecture)

Japan, and arsenic has long been used as a raw material for pesticides, rodenticides, wood preservatives. It was also used as a poisonous gas during the Second World War [26].

Arsenic pollution was caused by the production of arsenous acid several decades ago. The famous polluted areas are the Toroku and Matsuo mines in Miyazaki prefecture [26], the Sasagadani mine in Shimane prefecture, and the Tougane mine in Gifu prefecture. The Toroku and Matsuo mines in Miyazaki prefecture also manufactured arsenous acid using the same method. Arsenous acid is produced by refining arsenopyrite in the kiln [26] (Fig. 4.1). However, as no filter was provided to the kiln, arsenous acid was spread to the surroundings, and the environment was polluted by arsenic compounds. Therefore, many workers and residents around the mine suffered from arsenic poisoning.

Since Toroku became famous for patients suffering from arsenic pollution, the Matsuo mine without many patients was not as widely known as the Toroku mine. Mines produced gold, silver, and copper. From around 1920, the Matsuo mine produced arsenous acid from arsenopyrite in the same way as the Toroku mine. Former employees developed chronic arsenic poisoning, and they sued the management for damages to the mining industry. In 1983, the Miyazaki District Court ruled in favor of the plaintiffs [26]. The major difference between the Toroku mine and the Matsuo mine was that the Matsuo mine was located in a remote mountain area far from residential areas, and patients of arsenic poisoning were limited to former employees.

4.4 Arsenic Pollution in the Toroku Mine [26–28]

The Toroku mine was developed as a silver mine in the 1500s. The mine obtained its name from the technician invited over from Portugal. Toroku was a famous mine that produced silver under the direct control of the Edo Shogunate (1600–1868) and was lined up with the gold mine of Sado and the silver mine of Ikuno in Japan. In the Meiji era (1868–1912), Toroku mine produced tin, lead, copper, and zinc. Subsequently, arsenic-containing ores such as arsenopyrite were found in the Toroku mine, which began to produce arsenopyrite as a raw material for arsenous acid. Arsenous acid produced from the Toroku mine has been traded and exported as a raw material for herbicides, pesticides, and poisonous gas.

Arsenic is widely distributed in the crust and pollutes the environment through volcanoes and mines. Therefore, groundwater [29], vegetables [30], grain, and seafood [31] contain arsenic. Even the human body requires a very small amount of arsenic. In Toroku district, the environment was contaminated by arsenic trioxide, which was generated by burning arsenopyrite. This compound is also an anhydrous arsenous acid. When dissolved in water, it becomes arsenous acid ($As(OH)_3$). Arsenous acid is methylated in vivo and converted to organic arsenic. Organic arsenic is rapidly excreted in the urine. When arsenic exposure is low, it is excreted from the body through this metabolic pathway. However, if the exposure to arsenic is high, arsenous acid inhibits enzyme receptors with a thiol base [2]. In addition, arsenic can generate superoxide, which is an oxidative stress marker in the human body. It can also cause genotoxic and non-genotoxic effect [3]. Therefore, arsenic poisoning symptoms are mainly digestive symptoms, neuropsychiatric symptoms, liver dysfunction, and cardiovascular disorders [32]. If humans are chronically exposed to arsenic, skin diseases, skin cancer, and lung cancer could occur [33].

Patients with arsenic poisoning have already been reported in the Taisho era (1912–1926) around the Toroku mine. It has been reported that honey and shiitake mushrooms could not be harvested, and many cattle, horses, and residents died. In the Showa era (1926–1989), the Toroku mine produced more arsenous acid, and the number of patients suffering from arsenic poisoning increased. The management of the Toroku mine was transferred to the Teikoku Mining Development Company. They often conflicted with the residents, and the production of arsenous acid was occasionally interrupted. After that, the management rights of the Toroku mine were transferred to Nakajima Mining Co. Ltd., and production continued until 1937 when the mountain was closed. In the 1940s, Sumitomo Metal Mining Co. Ltd. managed and operated the Toroku mine.

To date, many cases of poisoning have been reported. However, this case of pollution had not been disclosed to the public by that time. The first research on this case of arsenic poisoning was conducted by a young teacher at an elementary school in Toroku district, who had just graduated from the Faculty of Education, Miyazaki University. First, the teacher wondered that students from Toroku district grew slowly and tended to fall sick. He also found that the teacher's wife, who is also from Toroku district, was prone to illness, referring to the records of his elementary school. Therefore, he wondered if there were health problems around the Toroku mine. In September 1969, the Miyazaki prefecture investigated the arsenic concentration of rivers. It was found that the wastewater from the Toroku mine contained a large amount of arsenic. However, they did not demonstrate an association between arsenic poisoning and patients' suffering.

Therefore, the teacher decided to start a survey on Toroku poisoning. In 1946, he proposed that the issues of Toroku should be taken up as an educational research theme for the workplace, and he set "poison and education" as the theme. The teachers started to explore the life history of the residents of Toroku district. As a result, he detected many health issues in residents and livestock and published the results. Since the pollution problem in Toroku district became widely known, in the same year, the Miyazaki prefecture also started a damage investigation and a health checkup for residents. The Miyazaki Prefecture indicated that there were diseases caused by arsenic exposure around the Toroku mine. Based on these results, the Miyazaki prefecture appealed to the Japanese government to treat these diseases as ones caused by arsenic pollution. On February 1, 1973, the Japanese government decided that Toroku district would be a "designated area for pollution-related diseases" and labeled these diseases as "chronic arsenic poisoning." The operation certified patients for chronic arsenism began in the polluted area according to the Pollution Health Damage Compensation Law, and 145 patients were certified by 1989 [34].

The production of arsenous acid continued from 1920 to 1962, with some interruptions. Since Toroku district was located in a V-shaped valley, polluted smoke was easily stagnant and close to the residential area, the living environment was polluted with a large amount of arsenous acid. The Miyazaki prefecture certified 108 men and 94 women (total 202) as patients with arsenic poisoning until March 31, 2017.

4.5 Review Related to the Toroku Mine

There were few articles on the Toroku mine [34–43], and some were written by Japanese authors [35–37, 39, 40, 42]. Tanaka compiled detailed data and published a book on arsenic pollution in the Toroku area [28]. Ishinishi et al. evaluated former workers of the Matsuo mine in Miyazaki prefecture [43]. However, they did not indicate the characteristic findings of arsenic poisoning, except gastrointestinal disturbance, cardiovascular disorders, hematopoietic disorders, and liver dysfunction. However, Tuda et al. (1990) indicated that analyses near the mine and refinery revealed high arsenic contents in the dust from ceiling boards of residences (200–8000 ppm), neighboring soils (2760 ppm on average), and percolated water from the slag (180 ppm) [36]. They also showed that cancer of the trachea, bronchus, lungs, respiratory and intrathoracic cancer, bladder, kidneys, and other unspecified urinary organs and ischemic heart disease significantly exceeded the expected value based on the mortality rate in Japan. When classified by employment history, the incidence of respiratory and urinary tract cancers among former miners was very high. In this way, there were many victims of arsenic poisoning in Toroku mine.

4.6 Current Environment Around the Toroku Mine

Toroku returned to the lush environment as shown in the photo (Fig. 4.2), and it is difficult to recall the time when the plants all withered in this area. The Miyazaki prefecture conducted a field survey around the Toroku mine. They shielded the inside of the mine with concrete to prevent the outflow of arsenic, and replaced

Fig. 4.2 Current landscape of Toroku district. It is a V-shaped valley

dangerous contaminated soil on 13 hectares with noncontaminated soil. The arsenic concentration in the river near the mine was still higher than the environmental standards. Therefore, they are not suitable for drinking water. However, the concentration of farm products is under standard and safe conditions.

Annual health checks to assess the effect of arsenic have been conducted every year by the Miyazaki prefecture, supported by the Miyazaki University Faculty of Medicine since 1973. The medical examinations included blood tests, internal and neurological examinations, dermatological examinations, and otorhinolaryngological examinations. To date, 4000 residents have been examined. In addition, new arsenic poisoning patients are still detected from health observations. The Toroku pollution, like other cases of pollution, made us reflect strongly on the fact that past mistakes leave a huge negative legacy in the future.

Acknowledgement The author gratefully appreciates participants of Miyazaki prefecture for providing the data and photo concerning to Toroku mine.

References

1. Sakurai T. Biomethylation of arsenic is essentially Detoxicating event. J Health Sci. 2003;49:171–8.
2. Hughes MF. Arsenic toxicity and potential mechanisms of action. Toxicol Lett. 2002;133:1–16.
3. Rossman TG. Mechanism of arsenic carcinogenesis: an integrated approach. Mutat Res. 2003;533:37–65.
4. Berg M, et al. Magnitude of arsenic pollution in the Mekong and Red River deltas--Cambodia and Vietnam. Sci Total Environ. 2007;372:413–25.
5. Xie R, et al. Heavy coal combustion as the dominant source of particulate pollution in Taiyuan, China, corroborated by high concentrations of arsenic and selenium in PM10. Sci Total Environ. 2006;370:409–15.
6. Enrique Biagini R. Chronic arsenic water pollution in the republic of Argentina. Med Cutan Ibero Lat Am. 1975;3:423–32.

7. Cornejo-Ponce L, et al. Levels of total arsenic in edible fish and shellfish obtained from two coastal sectors of the Atacama Desert in the north of Chile: use of non-migratory marine species as bioindicators of sea environmental pollution. J Environ Sci Health. Part A, Toxic/hazardous substances & environmental engineering. 2011;46:1274–82.
8. Tareq SM, et al. Arsenic pollution in groundwater: a self-organizing complex geochemical process in the deltaic sedimentary environment, Bangladesh. Sci Total Environ. 2003;313:213–26.
9. Dittmar T. Hydrochemical processes controlling arsenic and heavy metal contamination in the Elqui river system (Chile). Sci Total Environ. 2004;325:193–207.
10. Blanes PS, et al. Natural contamination with arsenic and other trace elements in groundwater of the central-west region of Chaco, Argentina. J Environ Sci Health. Part A, Toxic/hazardous substances & environmental engineering. 2011;46:1197–206.
11. Huq ME, et al. High arsenic contamination and presence of other trace metals in drinking water of Kushtia district, Bangladesh. J Environ Manag. 2019;242:199–209.
12. Raessler M. The arsenic contamination of drinking and Groundwaters in Bangladesh: featuring biogeochemical aspects and implications on public health. Arch Environ Contam Toxicol. 2018;75:1–7.
13. Ahmad SA, et al. Arsenic contamination in groundwater in Bangladesh: implications and challenges for healthcare policy. Risk Manag Healthc Policy. 2018;11:251–61.
14. Guo X, et al. Arsenic contamination of groundwater and prevalence of arsenical dermatosis in the Hetao plain area, Inner Mongolia, China. Mol Cell Biochem. 2001;222:137–40.
15. Karn SK. Arsenic (as) contamination: a major risk factor in Xinjiang Uyghur autonomous region of China. Environ Pollut. 2015;207:434–5.
16. Yu G, et al. Health effects of exposure to natural arsenic in groundwater and coal in China: an overview of occurrence. Environ Health Perspect. 2007;115:636–42.
17. Bacquart T, et al. Multiple inorganic toxic substances contaminating the groundwater of Myingyan township, Myanmar: arsenic, manganese, fluoride, iron, and uranium. Sci Total Environ. 2015;517:232–45.
18. Mar Wai K, et al. Arsenic exposure through drinking water and oxidative stress status: a cross-sectional study in the Ayeyarwady region, Myanmar. J Trace Elem Med Biol. 2019;54:103–9.
19. Ngoc NTM, et al. Chromium, cadmium, Lead, and arsenic concentrations in water, vegetables, and seafood consumed in a coastal area in northern Vietnam. Environ Health Insights. 2020;14:1178630220921410.
20. Nguyen VT, et al. Arsenic-contaminated groundwater and its potential health risk: a case study in LongAn and Tien Giang provinces of the Mekong Delta. Environ Sci Pollut Res Int: Vietnam; 2020.
21. Ruangwises S, et al. Dietary intake of total and inorganic arsenic by adults in arsenic-contaminated area of Ron Phibun district, Thailand. Bull Environ Contam Toxicol. 2010;84:274–7.
22. Yokoi K, et al. Toxicity of so-called edible hijiki seaweed (Sargassum fusiforme) containing inorganic arsenic. Regul Toxicol Pharmacol. 2012;63:291–7.
23. Ichikawa S, et al. Ingestion and excretion of arsenic compounds present in edible brown algae, Hijikia fusiforme, by mice. Food Chem Toxicol: An International Journal Published for the British Industrial Biological Research Association. 2010;48:465–9.
24. Nakamura Y, et al. Cancer risk to Japanese population from the consumption of inorganic arsenic in cooked hijiki. J Agric Food Chem. 2008;56:2536–40.
25. Mise N, et al. Hijiki seaweed consumption elevates levels of inorganic arsenic intake in Japanese children and pregnant women. Food Addit Contam Part A Chem Anal Control Expo Risk Assess. 2019;36:84–95.
26. http://toroku-museum.com/.
27. https://www.asia-arsenic.jp/.
28. Tanaka T Koudoku Toroku Jiken. Books Sanseido.
29. Das D, et al. Arsenic contamination in groundwater in six districts of West Bengal, India: the biggest arsenic calamity in the world. Analyst. 1994;119:168N–70N.

30. Datta DV, et al. Arsenic content of vegetables and soil in northern India--a concept of arsenicosis. J Assoc Physicians India. 1978;26:395–8.
31. Zook EG, et al. National Marine Fisheries Service preliminary survey of selected seafoods for mercury, lead, cadmium, chromium, and arsenic content. J Agric Food Chem. 1976;24:47–53.
32. Hutton JT, et al. Sources, symptoms, and signs of arsenic poisoning. J Fam Pract. 1983;17:423–6.
33. Sanderson KV. Arsenic and skin Cancer. Trans St Johns Hosp Dermatol Soc. 1963;49:115–22.
34. Tsuda T, et al. An epidemiological study on cancer in certified arsenic poisoning patients in Toroku. Ind Health. 1990;28:53–62.
35. Ohta T, et al. Studies on closed mines from the stand-point of social medicine. Report I. health hazards among workers and residents at Toroku mine (author's transl). Nihon eiseigaku zasshi. Japanese Journal of hygiene. 1976;31:457–64.
36. Hotta N, et al. A clinical study on Toroku disease-chronic arsenic poisoning due to environmental pollution. Bull Inst Constitutional Med Kumamoto Univ. 1979;3:199–235.
37. Hotta N, et al. A prognostic study on Toroku disease chronic arsenic poisoning due to environmental pollution. A 6 years follow up study. Bull Inst Constitutional Med Kumamoto Univ. 1984;3:559–76.
38. Tsuda T, et al. An epidemiological study on cancer in certified arsenic poisoning patients in Toroku. Sangyo Igaku. 1987;29:496–7.
39. Tsuda T, et al. A case of lung cancer associated with chronic arsenic poisoning caused by neighborhood exposure of As2O3 from Toroku mine. Sangyo Igaku. 1987;29:222–3.
40. Kawasaki S, et al. Chronic and predominantly sensory polyneuropathy in Toroku Valley where a mining company produced arsenic. Rinsho Shinkeigaku. 2002;42:504–11.
41. Ishii N, et al. Clinical symptoms, neurological signs, and electrophysiological findings in surviving residents with probable arsenic exposure in Toroku, Japan. Arch Environ Contam Toxicol. 2018;75:521–9.
42. Ohtsubo A. [pollution and chronic arsenic poisoning associated with the operation of former Toroku mine in Takachiho town, Miyazaki prefecture]. *Nihon eiseigaku zasshi*. Jpn J Hyg. 2018;73:275–6.
43. Ishinishi N, et al. Outbreak of chronic arsenic poisoning among retired workers from an arsenic mine in Japan. Environ Health Perspect. 1977;19:121–5.

Chapter 5
Yokkaichi Asthma: Health Effects of Air Pollutants in Japan

Kayo Ueda

Abstract Yokkaichi asthma from the 1960s to the early 1980s involved the onset and exacerbation of asthma, chronic bronchitis, and chronic obstructive lung disease, which was attributed to sulfur dioxide (SO_2). Inhaled SO_2 easily dissolves in the epithelial lung lining fluid of the nose and upper airways and generates secondary reactive compounds, such as sulfurous acid and sulfuric acid. These derivatives increase the level of prostaglandin D2, inducing the constriction of airway smooth muscle, inflammatory responses, and oxidative stress. As a result, exposure to SO_2 causes bronchoconstriction in asthmatic subjects and exacerbates asthma. Many epidemiological studies found that exposure to SO_2 was associated with mortality and incidence of asthma and chronic obstructive pulmonary diseases for adult, and prevalence of persistent cough/phlegm, asthma-like attack for schoolchildren. These scientific evidences played an important role to support the causal relationship between ambient SO_2 and chronic obstructive pulmonary disease in individual patients who filed a lawsuit against the petroleum industry group.

Keywords Air pollution · Health · Respiratory diseases · Sulfur dioxide · Asthma

5.1 Introduction

Yokkaichi asthma was one of the four major pollutant-based diseases in Japan during the high economic period, i.e., from the 1960s to the early 1980s. It involved the onset and exacerbation of asthma, chronic bronchitis, and chronic obstructive lung disease in Yokkaichi City that were attributed to the excessive emissions of sulfur

K. Ueda (✉)
Graduate School of Global Environmental Studies, Kyoto University, Kyoto, Japan
e-mail: uedak@health.env.kyoto-u.ac.jp

T. Nakajima et al. (eds.), *Overcoming Environmental Risks to Achieve Sustainable Development Goals*, Current Topics in Environmental Health and Preventive Medicine, https://doi.org/10.1007/978-981-16-6249-2_5

dioxide (SO_2). Serious health issues associated with similar air pollution were observed in other areas of the Pacific Belt from the Kanto plain to North Kyushu, where the coastal areas were occupied by oil refining and petrochemical industries [1]. Nine patients in the Isozu district filed a lawsuit against the companies operating the complex in Yokkaichi in 1967, and the plaintiffs won the case in 1972. Scientific evidence was used to establish a causal relationship between SO_2 and respiratory diseases. The following text describes the health effects of SO_2 based on both toxicological and epidemiological evidence.

5.2 Toxicological Evidence on the Health Effects of SO_2

The major sources of SO_2 include anthropogenic activities, especially the combustion of coal and oil containing sulfur, and metal smelting. Toxicological studies using human epithelial cells and animals have elucidated the mechanism by which exposure to SO_2 causes harmful effects on the respiratory system. Inhaled SO_2 easily dissolves in the epithelial lung lining fluid of the nose and upper airways and generates secondary reactive compounds, such as sulfurous acid (H_2SO_3) and sulfuric acid (H_2SO_4), which are subsequently converted to bisulfite and sulfite derivatives. These derivatives increase the level of prostaglandin D2, inducing the constriction of airway smooth muscle, inflammatory responses, and oxidative stress [2]. In addition, SO_2 oxidation contributes to the formation of secondary particles of various sizes, including fine and ultrafine particles. These secondary products and particles are also responsible for oxidative stress and inflammation in the respiratory tract [3].

Previous studies have shown that exposure to SO_2 alters mucociliary clearance, a defense mechanism of the respiratory system that removes inhaled particles and infectious agents. Sulfuric acid also affects mucociliary clearance and causes respiratory irritation [4]. Clinical and epidemiological studies have shown that asthmatics are more sensitive to SO_2. Genetic polymorphisms may enhance the susceptibility to respiratory effects of SO_2 [3]. A review of several controlled exposure studies, in which asthmatic volunteers were exposed to SO_2 during 5–10 min, has shown that exposure to SO_2 causes bronchoconstriction with increasing SO_2 between 0.2 and 1.0 ppm [5]. Another review reported that the response of lung function to SO_2 decreases in forced expiratory volume in the first second (FEV1), and an increase in airway resistance was induced within a few minutes. Although the concentration that causes acute responses varies according to each individual, exposure to SO_2 (>0.5 ppm) generally triggers bronchospasm in asthmatic patients [6].

5.3 History of Yokkaichi Asthma

After World War II, petroleum and petrochemical industries acquired the site of the former Japanese Navy Fuel Depot in the Yokkaichi area and developed petrochemical industry complexes (Table 5.1). A few years after its operation began in 1955,

Table 5.1 Timeline of air pollution-related events in Yokkaichi from World War II until 1973

Year	Events in Yokkaichi area	Events related to national and local government
1939–1945		World war II
		Petroleum and petrochemical industries started operation of its factory.
1955	Complaints about offensive odor in fish caught off Yokkaichi coast	Postwar high economic growth period in Japan policy for petroleum industry development by Ministry of International Trade and Industry. The petroleum industries used the ruins of the navy fuel deposit to build a petroleum complex.
1959		Several petroleum and petrochemical complexes started its operation.
Early 1960s	The complaints about noise, soot, malodor/irritating odor significantly increased in Shiohama complex areas	
1960	There was an increase in residences who complained asthma-like symptoms in Isotsu area.	Yokkaichi City pollution control committee was established. The monitoring of SO_2 (lead dioxide methods), and soot started.
1962		Soot and smoke regulation law was promulgated.
		Yokkaichi city pollution control committee reported that the level of soot was lower than Kawasaki city but SO_2 level was higher, especially in Isotsu area in the interim report.
	The health survey for residents was initiated.	Mie medical university carried out free health checkup to examine the possible health effects of air pollutants. They found a remarkably high prevalence of bronchial diseases.
1963		Yokkaichi pollution control council was established.
1965		Yokkaichi City started the certification system of pollution-related diseases. The Ministry of Health and Welfare initiated a health survey of air pollution effects on school children in Yokkaichi (until 1969).
1967	Nine patients in Isotsu filed a lawsuit against 6 companies.	The basic law for environmental pollution control was enacted. Several petroleum industries installed tall stacks.
1969	Several researchers testified on the causal association between air pollution and asthma at the trials.	The national government enacted the Act on Special Measures Concerning Relief of Pollution Related Disease.
1970	There were several public rallies against environmental pollution in Yokkaichi.	The national government started the support of medical expenses on the air pollution-related diseases.
1971		The environmental agency was established.

(continued)

Table 5.1 (continued)

Year	Events in Yokkaichi area	Events related to national and local government
1972	Judgment was issued. The court ordered the companies to pay compensation for the patients.	
1973		Environmental quality standards for SO_2 was established in Japan

This timeline was adapted from the table "the history of Yokkaichi pollution" with permission [in Japanese] (Yokkaichi Pollution and Environmental Museum for Future Awareness. https://www.city.yokkaichi.mie.jp/yokkaichikougai-kankyoumiraikan/pdf/pollution/nenpyo1-13.pdf. These are copyrighted materials)

there were claims about "petroleum odor" emanating from the fish caught near Yokkaichi harbor; this resulted in a huge economic loss for fisheries and was recognized as a social problem [7]. It was proved that aromatic hydrocarbons in seawater containing the petroleum industrial waste caused the offensive odor in fish [7, 8]. There were also resident complaints regarding noise, dust, and odors around the petroleum complex. In the 1960s, emissions of soot and sulfurous acid gases from the petroleum complex in Yokkaichi increased. Concurrently, a number of patients with asthma were reported in the Isozu district, the most polluted area, and its surrounding districts; this received wide public attention.

In response to the unusually high prevalence of asthma in these districts, a series of epidemiological studies have been conducted by national and local governments, and local universities [9]. From January to March 1963, the average SO_2 concentration measured by Thomas' volumetric method was 0.3 ppm, approximately 786 μg/m^3 in mass concentration, much higher than the WHO air quality guidelines [10] and the Japan Environmental Quality Standards [11], which are 20 μg/m^3 and 105 μg/m^3 for 24-h average, respectively, (Table 5.2). Moreover, the peak concentration often ranged from 1 to 2.5 ppm.

Studies using the National Health Insurance Billing data of over 10 districts with various SO_2 levels within the city found that the incidence of asthma and chronic obstructive pulmonary diseases for those aged 50 years and older was over 5 times higher in the polluted district than in the nonpolluted district [12]. A significant correlation was observed between the prevalence of bronchial asthma and SO_2 concentration. The correlation was evident among those aged 50 years and older [13]. Another study using data of Yokkaichi from 1961 to 1970 found that the mortality rate due to obstructive respiratory diseases, especially bronchial asthma and emphysema, was higher in polluted areas than in nonpolluted areas. The district-specific mortality rate due to obstructive respiratory diseases was strongly correlated with the concentration of sulfurous acid ($r = 0.76$ in 1971) [14]. A questionnaire survey on schoolchildren showed that the concentration of sulfurous acid was correlated with the district-specific prevalence of persistent cough/phlegm, asthma-like attack, and wheezing during a cold [15].

The evidence obtained from many epidemiological studies led by the local government and universities was used to support the causal relationship between

Table 5.2 Standards for SO_2 concentration in Japan and WHO

	Averaging time	Standards
Japan		
Environmental Quality Standards (1973)	24 h	105 μg/m^{3}[a]
	1-h	262 μg/m^{3}[a]
WHO		
Air quality guideline (2000)	24 h	20 μg/m^3
	10 min	500 μg/m^3
Interim target-2	24 h	50 μg/m^3
Interim target-1	24 h	125 μg/m^3

[a]Applying conversion factor of SO_2 for ppb to μg/m^3: 2.62
Source: Ministry of the Environment, Japan. Environmental quality standards in Japan - Air quality. Available from: https://www.env.go.jp/en/air/aq/aq.html
World Health Organization. WHO Air quality guidelines for particulate matter, ozone, nitrogen dioxide and sulfur dioxide. Global update. Global update 2005. Available from: https://apps.who.int/iris/bitstream/handle/10665/69477/WHO_SDE_PHE_OEH_06.02_eng.pdf

ambient SO_2 and chronic obstructive pulmonary disease (COPD) in individual patients who filed a lawsuit against the petroleum industry group [16]. After the plaintiffs won the case in 1972, Mie prefecture imposed total emission control of SO_2 for the entire Yokkaichi area based on the Mie Prefecture Ordinance that was more stringent than the national regulations [17]. SO_2 concentration decreased dramatically by one-third, 0.75 mg/100 cm^2/day with PbO_2 method in 1976 compared to 2.58 mg/100 cm^3/day in 1970. SO_2 is correlated with chronic obstructive lung disease, including bronchial asthma, chronic bronchitis, and emphysema, after accounting the area-level socioeconomic status [18].

Yokkaichi City was the first city in the country to establish a certificate system for pollution-related diseases in 1965. The city supported the medical expenses of patients who developed air pollution-related asthma, chronic bronchitis, and pulmonary emphysema regardless of the severity of the conditions [15]. Between 1965 and 1970, over 700 patients were acknowledged by this city-specific system. In 1969, the national government enacted the Act on Special Measures Concerning Relief of Pollution Related Disease and supported the certificate system. In 1987, the law changed to the Law Concerning Pollution-Related Health Damage Compensation and Other Measures.

Based on the statistics of 2018, over 2000 patients were officially designated as victims of Yokkaichi asthma. In 2018, 358 patients were said to have survived, although no new certified cases have been reported since 2012. Half of these cases included the elderly, aged 60 years and older. Approximately 80% of the patients were diagnosed with bronchial asthma, while the remaining were diagnosed with chronic bronchitis (Yokkaichi City Office, Report on environmental conservation, https://www.city.yokkaichi.lg.jp/www/contents/1543551934253/files/h30kankyouhozen.pdf).

5.4 Current Studies Related to Yokkaichi Asthma

Although many epidemiological studies conducted in the Yokkaichi area accumulated evidence supporting the adverse health effects of SO_2 on the respiratory system, most of them, especially those conducted before 2000 opted for an ecological study design because of limited access to individual data and rudimentary statistical methodologies. Since 2000, time-series analysis and case-crossover design have been used for air pollution epidemiology. These statistical methods account for confounders and have yielded more validated effect estimates of air pollutants. A study examined the association between SO_2 and mortality using daily data from 1972 to 1991 in Yokkaichi and its neighboring cities. The mean SO_2 level decreased dramatically from 32.9 ppb in 1972 to 7.8 ppb in 1991. A case-crossover design was applied in this study. After adjusting for relevant confounders, including seasonality, long-term trends, and meteorological factors, the results showed that an elevated level of SO_2 was associated with an increase in all-cause mortality [19]. The association between SO_2 and mortality was pronounced during the second half of the study period (1982–1991). In particular, the strongest association was observed between SO_2 and mortality due to COPD and asthma. The association continued even after adjusting for suspended particulate matter or nitrogen dioxide. Although an association between SO_2 and mortality from cardiovascular or cerebrovascular disease was also observed, it was not robust.

Air pollution has lifelong and acute effects. A study compared the mortality rate and life expectancy of 1354 patients who were certified as victims of air pollution-related diseases in Yokkaichi City during 1965–1988 with that of the entire population of Mie prefecture. Surprisingly, the mortality rate from COPD and asthma in the patients of Yokkaichi City was more than 10 times higher than that of the general population in Mie prefecture. Similarly, the life expectancy of patients in Yokkaichi was much shorter than that of the general population in Mie prefecture for any age group [20]. The average difference in the life expectancy for each age group ranged from 1.9 to 8.5 years for males and 4.5 to 8.5 years for females. Moreover, the reduction in life expectancy was larger for the younger age groups.

5.5 Conclusion

The case of Yokkaichi asthma shows that scientific evidence plays an important role in demonstrating the causal relationship between SO_2 and respiratory diseases in polluted areas. Japan's experience in environmental pollution is a lesson that unsustainable development will engender a huge public health burden arising from environmental pollutants.

References

1. Sargent J. Industrial location in Japan since 1945. GeoJournal. 1980;4(3):205–14. https://doi.org/10.1007/BF00218577.
2. Li R, Meng Z, Xie J. Effects of sulfur dioxide derivatives on four asthma-related gene expressions in human bronchial epithelial cells. Toxicol Lett. 2007;175(1–3):71–81. https://doi.org/10.1016/j.toxlet.2007.09.011.
3. Reno AL, Brooks EG. Mechanisms of heightened airway sensitivity and responses to inhaled SO2 in asthmatic patients. Environ Health Insights. 2015;9(Suppl 1):13–25. https://doi.org/10.4137/EHI.S15671.
4. Wolff RK. Effects of airborne pollutants on mucociliary clearance. Environ Health Perspect. 1986;66:223–37. https://doi.org/10.1289/ehp.8666223.
5. Johns DO, Svendsgaard D, Linn WS. Analysis of the concentration-respiratory response among asthmatics following controlled short-term exposures to sulfur dioxide. Inhal Toxicol. 2010;22(14):1184–93. https://doi.org/10.3109/08958378.2010.535220.
6. Peden DB. Mechanisms of pollution-induced airway disease: in vivo studies Allergy 1997 Oct;52(Suppl 38). 37–44; discussion 57. http://doi.wiley.com/10.1111/j.1398-9995.1997.tb04869.x.
7. Yoshida K, Uezumi N. The problem of offensive-odor fish in the petrochemical zone. [Japanese]. SEIKATSU EISEI (J. Urban Living Heal Assoc.). 1961;5(4):130–6. https://doi.org/10.11468/seikatsueisei1957.5.130.
8. Ogata M, Miyake Y. Identification of substances in petroleum causing objectionable odour in fish. Water Res. 1973;7(10):1493–504. https://doi.org/10.1016/0043-1354(73)90121-8.
9. Yoshida K, Oshima H, Imai M. Air pollution in Yokkaichi area with special regards to the problem of "Yokkaichi-asthma.". Ind Health. 1964;2(2):87–94. https://doi.org/10.2486/indhealth.2.87.
10. Ministry of the Environment, Japan. Environmental quality standards in Japan – Air quality. Available from: https://www.env.go.jp/en/air/aq/aq.html.
11. World Health Organization. WHO Air quality guidelines for particulate matter, ozone, nitrogen dioxide and sulfur dioxide. Global update. Global update 2005. Available from: https://apps.who.int/iris/bitstream/handle/10665/69477/WHO_SDE_PHE_OEH_06.02_eng.pdf.
12. Imai M, Oshima H, Takatsuka Y, Yoshida K. On the Yokkaichi-asthma. [Japanese]. Nippon Eiseigaku Zasshi (Japanese J Hyg). 1967;22(2):323–35. https://doi.org/10.1265/jjh.22.323.
13. Oshima H, Imai M, Fujita N, Fukuta H, Yoshida K. Air pollution and morbidity in Yokkaichi area. [Japanese]. J Jpn Soc Air Pollut. 1966;1(1):36–45. https://doi.org/10.11298/taiki1966.1.36.
14. Oshima H, Imai M, Kawagishi T. Air pollution and mortality in Yokkaichi area. [Japanese]. Nippon Eiseigaku Zasshi (Japanese J Hyg). 1971;26(4):371–6. https://doi.org/10.1265/jjh.26.371.
15. Imai M, Oshima H, Kawagishi T, Banno K, Yoshida K, Kitabatake M. The epidemiological studies on the effects of air pollution on high school students in Yokkaichi. [Japanese]. J Jpn Soc Air Pollut. 1973;8(5):703–9. https://doi.org/10.11298/taiki1966.8.703.
16. Maeda K. Problems in implication of epidemiology submitted as evidence to the court of air pollution health damages trials. J Epidemiol. 2001;11(6):276–80. https://doi.org/10.2188/jea.11.276.
17. Bianchi A, Cruz W. Local approaches to environmental compliance: Japanese case studies and lessons for developing countries [internet]. Washington, DC. World Bank; 2005. Available from: https://openknowledge.worldbank.org/handle/10986/7344?locale-attribute=es.

18. Imai M, Yoshida K, Kasama K, Tomida Y. A change in air pollution and its influence on the human body in Yokkaichi City. [Japanese]. J Jpn Soc Air Pollut. 1978;13(8):305–9. https://doi.org/10.11298/taiki1978.13.305.
19. Yorifuji T, Kashima S, Suryadhi MAH, Abudureyimu K. Acute exposure to sulfur dioxide and mortality: historical data from Yokkaichi. Japan Arch Environ Occup Heal. 2019 Sep 3;74(5):271–8. https://doi.org/10.1080/19338244.2018.1434474.
20. Guo P, Yokoyama K, Suenaga M, Kida H. Mortality and life expectancy of Yokkaichi asthma patients, Japan: late effects of air pollution in 1960–70s. Environ Health. 2008;7:8. https://doi.org/10.1186/1476-069X-7-8.

Chapter 6
"YOKKAICHI Studies" Learned from the YOKKAICHI Air Pollution for Environmental Policy and International Environmental Cooperation in Asia

Hye-Sook Park

Abstract "YOKKAICHI Studies" is a study that can be approached by four (4) aspects as follows: Firstly, it is a "Human Science" that investigates ownership of the dignity of life and nature—implying that the YOKKAICHI Air Pollution is not a problem that had already been solved, but rather an ongoing environmental issue. Secondly, it is a "Sustainable Futurology" that proposes a sustainable social system that strikes the balance between the environment and the economy in order to transform the former polluted YOKKAICHI City into a future environmental conservation city. Thirdly, it serves as a tool or blueprint for future generations (those who did not experience the Yokkaichi pollution) under the auspices of "UNESCO ESD-UNSDGs" to employ as a benchmark to develop global human resources. And lastly, it emphasizes the subject of "Environment and Asia" in the twenty-first century. In other words, in anticipating the occurrence of pollutions and other environmental issues in large industrial complexes in Asian countries during this era, the YOKKAICHI Studies was therefore defined as "Asian Studies"—dedicated to international environmental cooperation. To solve the transboundary air pollutions, such as yellow dust and PM2.5, it is expected of the leadership of Asia to demonstrate their know-how-based knowledge and determination to overcome among others the Yokkaichi air pollution and other environmental issues by engaging in an international environmental cooperation for action.

Keywords YOKKAICHI Studies · YOKKAICHI Air Pollution · ESD-SDGs Globalization · International Environmental Cooperation in Asia

H.-S. Park (✉)
Appointed Vice President for Environment & SDGs/Graduate School of Regional Innovation Studies of Mie University, Tsu, Japan
e-mail: park@mie-u.ac.jp

T. Nakajima et al. (eds.), *Overcoming Environmental Risks to Achieve Sustainable Development Goals*, Current Topics in Environmental Health and Preventive Medicine, https://doi.org/10.1007/978-981-16-6249-2_6

6.1 "YOKKAICHI Studies" Learned from the YOKKAICHI Air Pollution

"YOKKAICHI Studies" is a cross-disciplinary and synthetic environmental study learned from the YOKKAICHI Air Pollution. It was established to understand the past event of YOKKAICHI Air Pollution, to review the present and to propose a future image of YOKKAICHI City. Although the YOKKAICHI Industrial Complex built in the 1960s supported Japan's high economic growth period during the 1970s, it did cause heavy air pollution and YOKKAICHI Asthma, sacrificed the lives of innocent residents, increased water pollution of Ise Bay, and damaged the ecosystem of marine and the land. Besides, YOKKAICHI Studies is also an environmental pedagogy. It is an effective tool for Education for Sustainable Development (ESD) advocated by UNESCO, with the purpose of developing glocal (global-local) human resources that are rooted in the local area and also applicable to the world.

In addition, YOKKAICHI Studies is in harmony with environment, economy, and society, and it is based on the United Nations Sustainable Development Goals (UNSDGs)—the seventeen (17) goals adopted at the UN Summit in September, 2015 that need to address global issues by 2030. It is also a platform to provide scientific knowledge for sustainable regional revitalization.

"YOKKAICHI Studies" is a study that can be approached by four (4) aspects as follows: Firstly, it is a "Human Science" that investigates ownership of the dignity of life and nature—implying that the YOKKAICHI Air Pollution is not a problem that had already been solved, but rather an ongoing environmental issue. Secondly, it is a "Sustainable Futurology" that proposes a sustainable social system that strikes the balance between the environment and the economy in order to transform the former polluted YOKKAICHI City into a future environmental conservation city. Thirdly, it serves as a tool or blueprint for future generations (those who did not experience the Yokkaichi pollution) under the auspices of "UNESCO ESD-UNSDGs" to employ as a benchmark to develop global human resources. And lastly, it emphasizes the subject of "Environment and Asia" in the twenty-first century. In other words, in anticipating the occurrence of pollutions and other environmental issues in large industrial complexes in Asian countries during this era, the YOKKAICHI Studies was therefore defined as "Asian Studies"—dedicated to international environmental cooperation. To solve the transboundary air pollutions, such as yellow dust and PM2.5 (Fig. 6.1).

6.2 Litigation Judgement of YOKKAICHI Air Pollution and Environmental Policies by Mie Prefecture and YOKKAICHI City

Starting from Ashio Copper Mine Mineral Pollution Incident in the 1870s, then people have experienced the Minamata disease, Itai-Itai disease, Niigata Minamata disease, and YOKKAICHI Asthma in the 1960s.

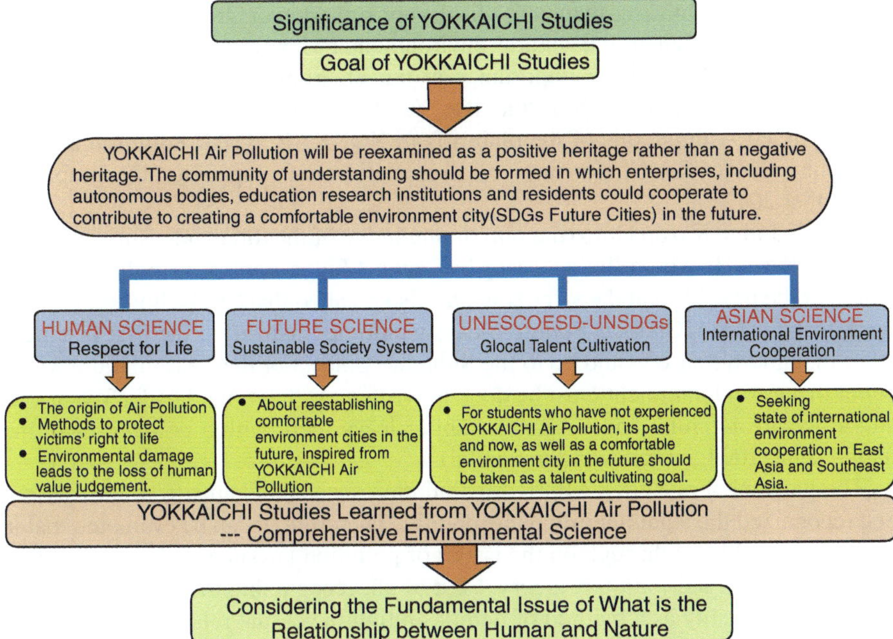

Fig. 6.1 YOKKAICHI Studies learned from YOKKAICHI Air Pollution

Air pollution in YOKKAICHI reached its worst levels around 1963–1964, as heavy oil sulfur content rates measured 3% more or less, annual sulfur dioxide estimated at 130,000–140,000 tons was polluted. Hourly measurements of sulfur dioxide concentrations in the Isozu (which was the nearest district to YOKKAICHI petrochemical complex in 1964) showed that the number of values exceeding 0.5 ppm amounted, occasionally rising above 1 ppm (more than 10 times the current environmental quality standard of 0.1 ppm). The annual average sulfur dioxide concentration in this area was 0.075 ppm (almost a 4 times more than the current environmental quality standards).

A pollution victim compensation system was introduced; which later came to specify YOKKAICHI Asthma as including such diseases as chronic bronchitis, bronchial asthma, and pulmonary emphysema. Statistical data indicates that after the compensation measures were introduced in the 1970s, the number of certified victims who suffered in YOKKAICHI City totaled 1738 with a peak number of 1140. The majority of victims of the YOKKAICHI Asthma were obviously children, aged people, and women who were put under weak health conditions in both physiological and social aspects.

Because of relying too much on science and technology, prioritizing national policies and corporates' profit pursuit and lacking awareness in protecting vulnerable residents, it is necessary to establish values that are based on environmental justice and to create a sustainable society.

"YOKKAICHI City was eventually damaged because of the YOKKAICHI industrial complex pollution. The day of July 24th, 1972, the day that judgement was issued, I couldn't say thanks to all of you. But when YOKKAICHI industrial complex got its original and natural environmental view as it should be, and all people really care about the importance of our environment, I would like to say thank you to all of you." Said the late Yukikazu Noda, a certified patient of YOKKAICHI Asthma, one of the plaintiffs of YOKKAICHI Air Pollution lawsuit (Autumn Academic Conference of Association of Japanese Geographers, September 2017).

Former Judge Kazuo Goto (the only living judge of the three judges in charge of the YOKKAICHI Air Pollution case) had said: "Judge applies to judgment sentences of the trial." But at the same time, he also recalled that "To be honest, I didn't think that YOKKAICHI Air Pollution case would be finalized in the first instance, and I thought this case would go to the Supreme Court. For that reason, there was a belief that a solid judgment must be done or written. At that time, I had a 3-year-old and a 5-year-old child, and I did not want to leave them with a polluted environment" (June 2004, interviewing in Sendai).

The judgement of YOKKAICHI Air Pollution accepted claims from the victims and recognized the legal liability of corporates. Therefore, it can be evaluated that it was an epochal breakthrough on the issues of pollution lawsuits such as epidemiological causality and joint tort theory. First of all, recognizing the theory of epidemiological causality as a method of demonstrating the casual relationship between corporates' tortious actions and the damage of residents has great significance.

Instead of individual corporates, corporates group, which formed YOKKAICHI Industrial Complex, were recognized that they must be responsible for the damage compensation. In another word, the judgment has got its most successful achievement in establishing that responsibility is inevitable for results. After that, the regulation of total amount and Pollution-related Health Damage Compensate Law were enacted as environmental policies of Mie Prefecture and YOKKAICHI City since 1972. The concentration of air pollution at the YOKKAICHI Industrial Complex has not exceeded the national environmental standard values of sulfur dioxide concentrations (The daily average of the 1-h value shall be 0.04 ppm or less, and the 1-h value shall be 0.1 ppm or less) in compliance with the regulation of total amount of Mie Prefecture and YOKKAICHI City since 1972. With the clarification of corporates' responsibilities and implementation of strict pollution control, the judgment of YOKKAICHI Air Pollution became a turning point, which makes the pollution control measures of Japan escape from the formerly vague versions and turn into a legitimately adaptive and effective new version.

When the victims of pollution are standing in a weak position of society, it is hard to protect them by policies which prioritized general public interest. And regarding public interest, pollution issues are closely related to the status of the country. Through proceedings, the citizen governance of a mature civil society, establishing a triune system that combines not only fulfillment of corporate social responsibility (CSR), but also the creating shared value (CSV) with justified environmental policies to protect the ecosystem including human beings. We could expect the construction of YOKKAICHI City in the future, as one of the environmentally advanced cities.

6.3 Glocalization of Human Resource Development and Creating Sustainable City through Collaboration of ESD and SDGs

ESD was proposed by Japanese government during Johannesburg World Summit, August 2002 (the summit of sustainable development), and at the 2005 UN Assembly, the UN 10-year ESD international practice plan formulated by UNESCO, and incorporating principles, values, and practice of sustainable development in teaching and learning which was also approved at the same time. The goals of ESD are learning and activities that aim to create a sustainable society by helping people take global level issues as their own problems, starting from environmental problems near around and creating new values and actions while solving the environmental problems. As education of nurturing leaders in creating sustainable society, ESD also is a comprehensive learning approach that connects various related fields from the perspective of establishing sustainable society.

SDGs, conformably adopted during UN Sustainable Development Summit (September 2015), are international goals that will be addressed by all developing countries, emerging countries, and developed countries during 2016–2030. With the concept of "No one will be left behind," SDGs serve as an important guideline for the international community to achieve sustainable society and pursue the global partnership in which all stakeholders such as government, corporates, academies-schools, and citizens are cooperating. SDGs cite 5Ps—"People," "Planet," "Prosperity," "Peace," "Partnership" as important elements of sustainable development.

In order to achieve the ultimate goal of the seventeen (17) SDGs; which are characterized by equal access to opportunities and resources for all people so as to realize a sustainable society in which the environment, economy, and society are harmonized, the "YOKKAICHI Studies" (which draws lessons from the YOKKAICHI Air Pollution) will be the most effective tool for learning the SDGs. In particular, Goal 4 is about education—"Ensuring inclusive and equitable quality education and promote lifelong learning opportunities for all." The ring of ESD-SDGs is necessary in nurturing leaders for sustainable society.

6.4 International Environmental Cooperation in Asia

On the basis of 1970s' economic development plans, in Ulsan, Onsan, and Yeosu-its coastal areas, Korea built National Industrial Complexes based on petrochemical industry. And Korea has turned from a developing country into a developed country. Then, in the 1980s, typified by "Onsan Disease," composites of YOKKAICHI Asthma. The health hazards of asthma to local residents caused by air pollutants such as sulfur dioxide from the National Industrial Complexes in Korea are similar to the mechanism of occurrence of YOKKAICHI Asthma caused by air pollutants from the YOKKAICHI Industrial Complex.

However, Korean government did not admit that there were pollutions and environmental problems, continuing the operation in national industrial complex. And in the 2000s, the government attempted to solve the problems by prompting mass migration of residents who were living in polluted areas. The residents of Ulsan and Onsan left their villages with the following words on a monument built in hometown. "From a long time ago, in this land children and grandchildren have lived in harmony with each other, now the leaving time has come. With our backs to this land and this sea, we must leave. Walking away from this land, will our ancestors forgive us and still protect us? We resign our painful feelings to this monument, depart for a faraway place. From now on, we can only beg that this land would be blessed and could achieve further development"—April 2001, Yongyeon Local Community.

A study by Park (2004) investigating the correlation between residential areas and asthma, chosen about 1000 elementary school students around YOKKAICHI Industrial Complex and about 2000 elementary school students around Ulsan, Onsan, Yeosu National Industrial Complex in Korea as research subjects, has showed that in both countries, the proportion of children with asthma tended to be high near the industrial complex. Especially about 30% of children who live around the industrial complex in Korea showed asthma symptoms, just like the case of YOKKAICHI Asthma. And it was clear that there is an urgent need to deal with pediatric asthma.

Transboundary air pollutions, such as yellow dust and PM2.5 which along with industrial activities and rapid increase of automobiles, not only causing economic damages and healthy damages in China, but also influencing Korea and Japan due to the westerlies. The international environmental problems are becoming more and more apparent.

First, desertification due to global warming is becoming more prominent, and the source of yellow dust is moving eastward from the Gobi Desert in Mongolia to Inner Mongolia and northeastern China, resulting in a noticeable increase in the number of warning days due to yellow dust. The average number of days of yellow dust warnings in Korea has increased rapidly from about 4 days in the 1980s to about 8 days in the 1990s and about 13 days in the 2000s. This phenomenon is also manifesting itself in Japan.

Next, with regard to PM2.5, the most important issue is the health hazards caused by transboundary air pollution from China, Korea, and Japan, as the effects of economic growth in China.

To solve the international environmental problems in Asia, we can expect that a leadership, which come from the know-how in overcoming YOKKAICHI Air Pollution and the form of international environmental cooperation regime, will take an action. As a win-win strategy for international environmental cooperation against transboundary air pollution such as yellow dust and PM2.5 in Japan, China, and Korea, the following strategy should be considered based on YOKKAICHI Studies, taking advantage of the lessons learned from YOKKAICHI Air Pollution.

1. Establishment of an International Environmental Cooperation Platform.

   At the government level, the monitoring system led by the East Asia Network for Acid Rain Monitoring (EANET) should be operated and information under the initiative of Japan since the 1990s.
2. Establishment of an International Environmental Research Network of Universities and Research Institutes as an Epistemic Community.

   Universities and research institutes are required to share their scientific knowledge by operating an international environmental research network to solve the problem of yellow dust and PM2.5 in Asia. In addition, international environmental activities by researchers and students from China, Korea, and Japan are required through the operation of the ESD-SDGs International Network for the development of international environmental human resources.
3. Establishment of an International Environmental NGOs Network.

   There is a need to raise public awareness of the global environment through private-sector exchanges among China, Korea, and Japan.

In August 1995, AANEA (Atmosphere Action Network East Asia) has been established with the aim of realizing YOKKAICHI Studies to learn from YOKKAICHI Air Pollution, by Environmental NGOs from seven countries/region to address these problems including China, Hong Kong, Japan, Mongolia, the Russian Far East, Korea, and Taiwan. The AANEA was founded under the leadership of Park and other members and Park is the representative of AANEA.

Since 2000, the name has been changed from AANEA to AANEA/EANEA (Environmental Action Network East Asia) because it concerns not only the atmospheric environment but also the environment in general.

References

1. Ueno T, Park H-S, editors. Aiming for the environmentally advanced city: Environmental science for open the future. Tokyo: Chuouhouki Publisher; 2004.
2. Park H-S, editor. YOKKAICHI studies: Recommendations from YOKKAICHI air pollution. Nagoya: Fubaisya Publisher; 2005.
3. Park H-S, editor. YOKKAICHI studies lecture. Nagoya: Fubaisya Publisher; 2007.
4. Park H-S, editor. Challenge of YOKKAICHI studies to question the past, present and future of YOKKAICHI air pollution. Nagoya: Fubaisya Publisher; 2012.
5. Park H-S, editor. Mie studies. Nagoya: Fubaisya Publisher; 2017.

Chapter 7
What we Have Learned from the Pollution that Occurred in Japan a Half Century Ago

Keiko Nohara, Suminori Akiba, Mayumi Ishizuka, and Tamie Nakajima

Abstract The development of industry enriches people's lives, yet on the other hand, it can cause serious disasters if not properly managed. From the early 1900s, Japan experienced several major pollution incidents due to severe environmental disruption by industrial activities. Those incidents have damaged human life and health. It was not until the 1960s that pollution-related diseases were officially designated. In February 2019, about half a century after the designation, we held a public symposium to review the history of pollution incidents that have occurred in our country and to discuss what we think about environmental issues today. In this paper, we introduce the lectures and discussions at the symposium, and reconsider what we have learned from those pollution incidents.

Keywords Pollution · Environmental contamination · Occurrence of pollution Response to pollution incidents · End of contamination · Basic Law for Environmental Pollution Control

K. Nohara (✉)
Center for Health and Environmental Risk Research, National Institute for Environmental Studies, Tsukuba, Ibaraki, Japan
e-mail: keikon@nies.go.jp

S. Akiba
Hirosaki University, Hirosaki, Japan
e-mail: k9242954@kadai.jp

M. Ishizuka
Faculty of Veterinary Medicine, Hokkaido University, Sapporo, Japan
e-mail: ishizum@vetmed.hokudai.ac.jp

T. Nakajima
College of Life and Health Sciences, Chubu University, Kasugai, Aichi, Japan
e-mail: tnasu23@isc.chubu.ac.jp

T. Nakajima et al. (eds.), *Overcoming Environmental Risks to Achieve Sustainable Development Goals*, Current Topics in Environmental Health and Preventive Medicine, https://doi.org/10.1007/978-981-16-6249-2_7

7.1 Introduction

In the mid-1900s, Japan experienced serious environmental pollution (Kohgai in Japanese) in many areas in the country. About 50 years after the recognition of individual pollution-related diseases by the government, on February 3, 2019, the Science Council of Japan (SCJ) and the Japanese Society for Hygiene (JSH) held a joint public symposium, "What We Think Now, Half a Century after the Recognition of Pollution-related Diseases: Toward the Achievement of the Sustainable Development Goals," at the 89th Annual Meeting of the JSH [1]. At that symposium, five speakers gave lectures: "Minamata Disease" (speaker, Murata), "Itai-itai Disease" (Aoshima), "Toroku Arsenic Poisoning" (Kuroda), "From Pollution Trials to Future Action Goals" (Otsuka), and "Efforts of Environmental Model Cities" (Nakamura). The articles by each speaker can be found in Part 1 and Part 3 of this book. After the lecture, a discussion session was held to take comments from the audience.

This paper reviews the historical pollution incidents from their inception to end, and reconsiders what we have learned, and what we think about them today.

7.2 Occurrence of the Pollution

The major pollution incidents that occurred in our country include Minamata disease, *Itai-itai* disease, and Toroku arsenic poisoning, which were discussed in the symposium mentioned above, as well as Niigata Minamata disease and Yokkaichi asthma. All cases were due to the release of wastewater and/or soot containing harmful chemical substances into the environment.

Minamata, Toroku, and Yokkaichi are the names of the areas where these pollution incidents occurred (Fig. 1.1 in Chap.1). The causal chemical substance of Minamata disease in Kumamoto Prefecture was methyl mercury in wastewater discharged into Minamata Bay from a factory (Murata, Part 1, Chap. 2). In Miyazaki Prefecture, Toroku arsenic poisoning was caused by effluent and exhaust gas containing arsenic from mines (Kuroda, Part 1, Chap. 4). Yokkaichi asthma was caused by exhaust gas from industrial complexes in the port area of Yokkaichi, Mie Prefecture (Ueda, Part 1, Chap. 5; Park, Part 1, Chap. 6). Niigata Minamata disease was pollution that occurred in Niigata Prefecture (Fig. 1.1 in Chap. 1), and the cause was methyl mercury, the same as Minamata disease. Itai-itai disease derives its name from "it hurts" ("*itai*" in Japanese), since this disease brought intense pain. Itai-itai disease was caused by drainage containing cadmium from mines (Aoshima, Part 1, Chap. 3).

Toroku arsenic poisoning and Itai-itai disease were already reported in the early 1900s. Then, in the mid-1950s, Japan entered a period of rapid economic growth after the defeat in World War II in 1945. In 1955–1973, the average annual economic growth rate exceeded 10%. During this period, industrial activities expanded greatly, but at the same time, many issues, such as air and water pollution, the destruction of nature, and noise and vibration problems, were becoming more and more apparent and serious in many areas in Japan.

In the Minamata area, for example, unusual behavior was observed in cats and crows which ate fish caught in Minamata Bay. Even so, for some time it was not generally thought that pollution could cause damage to humans. Even after the damage to people became apparent, there was an idea of economic priority in society, and human life and health were not considered the top priority. One of the examples showing such thought was the harmony clause of the Basic Law for Environmental Pollution Control enacted in 1967, as described later. The situation led to the occurrence of severe pollution in various parts in Japan that continued until early 1970s.

7.3 Response to the Damage

Taking Minamata disease as an example, the first official report was made by physicians of the responsible company who reported cases of the disease to the Minamata Public Health Center in 1956. In response, physicians and researchers from the local government, Kumamoto University, and national government began to investigate the cause of the disease. The details are described elsewhere [2, 3]. The records show that experts worked hard to carry out research to identify the cause of the disease as described below. However, it was not until 12 years later, in 1968, that the government admitted methyl mercury which contaminated the wastewater discharged from a factory to Minamata Bay as the causative agent. As for the fact that it took a long time to determine the cause, it was cited that the responsible company did not cooperate with the investigation and continued to refute external surveys. Among other reasons were that the research team did not include researchers who were familiar with chemical reactions, and that measuring mercury was technically difficult at that time. In the meantime, the damage spread as appropriate measures were not taken.

In the Minamata disease trial, important judgments on the responsibility of the company and that of the government and the prefecture were issued in 1973 and 2004, respectively (Otsuka, Part 3, Chap. 15). The judgement in 1973 found the company's negligence in not investigating their own emissions, even at the stage where the causal substance had not been identified. The 2004 judgement ruled that the government's failure to exercise its regulatory authority was illegal. For example, the failure of the central government meant the failure of the Minister of Trade and Industry to exercise his authority to designate Minamata Bay as the designated area based on the Public Water Area Quality Conservation Act and the Factory Discharge Regulation Act. These judgements share the same concept as "the precautionary principle." The principle has been advocated internationally in environmental issues since the 1980s. It intends that when serious damage to humans or the environment is predicted by certain scientific data, the events need to be controlled to prevent the damage from spreading, even if the cause is not completely proven scientifically.

On the other hand, another difficult issue was patient certification, which was based on the Law concerning the Relief of Pollution-related Health Damage. In Minamata disease, there were many factors which delayed certification, such as strict criteria, a lack of enough knowledge about specific symptoms, and a lack of appropriate measurement methods of causative agents (Murata, Part 1, Chap. 2). Not only for Minamata disease but also for other pollution diseases, patient certifications are still current issues that need to be resolved.

7.4 End of Severe Contamination

In the 1960s, environmental pollution caused by industrial activity increased in many parts of Japan, and residents' movements against the destruction of the environment became more active [4]. In response to this growing public opinion, the government enacted a basic law that established a framework for the promotion of anti-pollution measures, the Basic Law for Environmental Pollution Control of 1967. It states that "the purpose of this law is to promote comprehensive measures against pollution, thereby protecting the health of the people and preserving the living environment." However, the clause was followed by the following harmony clause: "The preservation of the living environment shall be harmonized with the sound development of the economy."

Three years later, in response to public opinion and other opposition, the harmonization clause was removed in 1970. Finally, this law as well as other related legislations clearly stated the idea that human health should take precedence over everything else. Another statement was that the air, sea, and rivers should not be polluted. In 1971, the government established the Environment Agency to centralize responsibility for pollution control, nature conservation, and planning and coordination of environmental policies. At the United Nations Conference on the Human Environment in 1972, Japanese government representatives expressed regret over the occurrence of severe pollution in Japan and their determination to prevent environmental destruction. With these policy changes since 1970, the pollution caused by chemicals released into environment came to an end.

7.5 Citizens' Initiatives for Environmental Issues

As described in Nakamura's article (Part 3, Chap. 16) in this book, in areas where pollution occurred, there was not only enormous damage to human health and the environment, but also conflicts and discrimination between patients, local residents, and residents of other areas. Entire communities are severely burdened by pollution incidents. By learning the lessons from those experiences, Minamata City announced the declaration of "Creating an Environmental Model City" in 1992. This activity has continued to the present.

At the above-mentioned joint symposium, a question was asked about the training of interpreters who participate in environmental conservation, share problems with local people, and transfer them to other areas. As a reply, it was noted that efforts are being made in Minamata City to pass on the past experiences to the third generation who have not experienced pollution. It is also important for citizens to be conscious of the environment and consider it together with the government.

7.6 Current and Future Challenges of Health Effects from the Environment

After the pollution incidents were contained, the quality of air and water in Japan has been improved. The focus of environmental health issues has shifted to low dose issues. At the joint symposium mentioned above, the audience raised an issue of how to present adverse health effects of a contaminant that is of local and natural origin, exists at a low level, and is deeply involved in daily life. To deal with these problems, there will be an increasing need to discuss the interpretation of data and the risk communication that informs it.

Recent studies have pointed out a variety of emerging concerns on health effects of the environment. For example, gestational exposure to environmental factors is reported to cause not only the effects seen at birth, but also those that appear in adulthood and/or in later generations. Minamata disease was one of the first cases which showed that fetuses are sensitive to environmental insults. It was one of many lessons we have learned. Until we know, we cannot prevent. Researchers need to conduct steady research with foresight and without preconception. To have foresight, we need to learn.

The Basic Law for Environmental Pollution Control, which was revised in 1970, was succeeded by the Basic Environment Law in 1993. The Chap. 1, Article 1 states that the purpose of the law is "to contribute to ensuring a healthy and cultured life for the people of the present and future, and to contribute to the welfare of humankind." We are responsible not only for the health of ourselves but also that of future generations. To fulfill that responsibility, we have been learning much from our past improper conduct.

References

1. Joint symposium of the science council and the hygienic society of Japan. Japanese J Hygiene, 74:S77–82, 2019.
2. Murata K, et al. Minamata disease. Encyclopedia of Environmental Health, vol. 3. Burlington: Elsevier; 2011. p. 774.
3. Social Science Study Group on Minamata Disease. Not to repeat the tragedy of Minamata Disease: what we can learn from the experience of Minamata Disease. Minamata: National Research Center for Minamata Disease; 1999. (In Japanese)
4. Shoji H, Miyamoto K. Pollution in Japan. Tokyo: Iwanami Shoten Publishers; 1975. (In Japanese)

Parts II
Environmental Issues That the World Is Watching Now

Chapter 8
Current Status and Problems Concerning the Management of Persistent Organic Chemicals Used in Products and those of Unintentional Products in Japan

Shigeki Masunaga

Abstract With the global initiatives for chemical management such as Stockholm Convention on Persistent Organic Pollutants (POPs), Agenda 21, and SAICM, the Government of Japan took some positive measures against PCBs wastes, industrial chemicals, and unintentionally formed POPs. Here, those measures and their results will be explained, and the problems left will be discussed. The POPs listed by Stockholm Convention are designated as class I specified chemical substance under the Chemical Substance Control Law in Japan and their production, import, and use are strictly regulated. The law, however, works only as follow-up actions though it has re-inspection system on existing chemicals. As for unintentionally formed POPs, their emission inventories are estimated and updated every year expecting the better management by related facilities.

Keywords Chemical Management · Persistent Organic Pollutants (POPs) Stockholm Convention · SAICM · PCBs · Dioxins · Unintentional Products

8.1 Introduction

Numerous synthesized chemical substances are used in consumer products such as electric appliances, furniture as well as food containers and packaging to support our convenient daily life. They also support food production as agrochemicals and fertilizers and healthy life as pharmaceuticals, preservatives, insecticides, and surfactants. Life without chemical substances is hard to imagine, however, we must

---

S. Masunaga (✉)
Yokohama National University, Yokohama, Japan
e-mail: masunaga-shigeki-dh@ynu.ac.jp

T. Nakajima et al. (eds.), *Overcoming Environmental Risks to Achieve Sustainable Development Goals*, Current Topics in Environmental Health and Preventive Medicine, https://doi.org/10.1007/978-981-16-6249-2_8

pay attention to their hazardous side because human as well as biota encounter those newly synthesized chemicals for the first time.

Japan experienced severe human health damage due to environmental pollution caused by industries, the so-called Kohgai (Public health hazard). By introducing strict regulations on limited number of hazardous chemicals, we succeeded to a certain extent in calming down Kohgai, and we have been working to manage a variety of chemicals that permeate into our daily lives. However, our effort is not necessarily leading the world, indeed there are cases in which we have been urged by the movement of the world. Here, I will discuss the progress and challenges in the management of chemical substances in Japan and to contemplate how we can better deal with the problem.

8.2 Progress in International Chemical Management

The first international move to environmental conservation is the United Nations Conference on the Human Environment in Stockholm, 1972. At the conference, under the slogan of "Only One Earth," the "Declaration of the United Nations Conference on the Human Environment" and the "Action Plan for the Human Environment" were adopted, and the United Nations Environment Programme (UNEP) was established for their implementation. In 1992, the United Nations Conference on Environment and Development (UNCED), the Earth Summit, was held in Brazil, and the "Rio Declaration" and its action program "Agenda 21" were adopted. Its Article 19 clarified the need for the management of chemical substances [1]. Ten years later in 2002, the World Summit on Sustainable Development (WSSD), the Johannesburg Summit, was held in South Africa, and the "Johannesburg Declaration on Sustainable Development" and "Johannesburg Plan of Implementation" were adopted, reviewing the Agenda 21. Regarding chemical substance management, it was agreed to "to achieve, by 2020, that chemicals are used and produced in ways that lead to the minimization of significant adverse effects on human health and the environment" (Paragraph 23, [2]). This is called the "2020 Goal" and it worked as a stimulus for many countries to reform their chemical management systems by 2020. In 2006, the International Conference on Chemicals Management was held in Dubai, and the "Strategic Approach to International Chemical Substance Management (SAICM)" was adopted to materialize the goal. It clearly described, "To ensure by 2020: that chemicals or chemical uses that pose an unreasonable and otherwise unmanageable risk to human health and the environment based on a science-based risk assessment and taking into account the costs and benefits as well as the availability of safer substitutes and their efficacy, are no longer produced or used for such uses; and that risks from unintended releases of chemicals … are minimized" [3].

To attain the goals described in SAICM, the European Union (EU) brought a strict chemical management system called "Registration, Evaluation, Authorisation and Restriction of Chemicals (REACH)" into force in 2007, requiring the manufacturers and importers to assess risk of chemicals all through their supply chain. In Japan, the Act on the Evaluation of Chemical Substances and Regulation of Their Manufacture, etc. (Chemical Substances Control Law) was revised in 2009, and the assessment method of chemical substances was converted from hazard based to risk based assessment. Japanese government started to re-inspect the existing chemical substances. In the United States, the Chemical Substances Control Act (TSCA) was first amended in 2016 to strengthen the authority of the Environmental Protection Agency and to establish the relevant laws. In this way, countries have made efforts to achieve the 2020 goal.

Coming close to the year 2020, the UNEP published the "Global Chemicals Outlook II" on March 11, 2019. This second report summarized "The global goal to minimize adverse impacts of chemicals and waste will not be achieved by 2020. Solutions exist, but more ambitious worldwide action by all stakeholders is urgently required" [4]. While the size of the global chemical industry is projected to double by 2030, progress in reducing chemical risk is not catching up. Further actions are needed especially in developing countries.

8.3 International Move in Persistent Organic Pollutants Management

Here, I take persistent organic pollutants (POPs) as an example to show the international progress of chemical regulation. POPs are defined as chemicals that have the following four properties, namely, adverse effect, persistence in the environment, bioaccumulation, and potential for long-range environmental transport. As POPs can be a threat to human health and to the environment, "The Stockholm Convention on POPs" was adopted in 2001 and entered into force in 2004. The convention specifies POPs in three categories: (A) prohibition of manufacturing, use, import and export, (B) restrictions on manufacturing, use, import and export, and (C) reduction of emissions of unintentional products, and requests the binding countries to appropriately regulate them and to destruct their stock and waste. For polychlorinated biphenyls (PCBs) in particular, the countries are imposed a duty of effort to abolish its use by 2025 and to destruct its waste liquids and equipment by 2028.

At the birth of the convention, the number of POPs was only twelve and most of them were agrochemicals (pesticides). In 2018, the number of POPs grew up to 30 groups of chemicals. Chemicals that were used in daily necessities, furniture, and electric appliances have come to be designated (Fig. 8.1). This means that chemicals that had been circulated both domestically and globally would suddenly have to be regulated. Alternative chemicals must be developed to replace them and POPs containing consumer products must be collected and disposed of properly.

Pesticides
Aldrin (2001) A
Chlordane (2001) A
Dieldrin (2001) A
Endrin (2001) A
Heptachlor (2001) A
Mirex (2001) A
Toxaphene (2001) A
DDT (2001) B
Chlordecone (2009) A
Lindane (2009) A*
α-Hexachlorocyclohexane (2009) A
β-Hexachlorocyclohexane (2009) A
Technical endosulfan and its related isomers (2011) A*
Pentachlorophenol and its salts and esters (2015) A*
Dicofol (2019) A
Hexachlorobenzene (2001) A・C
Pentachlorobenzene (2009) A・C

Explanatory Notes
(Year): Year listed as POPs (Effective year may be later)
Regulation category
A: Eliminate the production and use
B: Restrict the production and use
C: Reduce the unintentional release
*: Exception in some certified uses.

Industrial Chemicals
Tetra- & Penta-bromo-DE (2009) A*
Hexa- & Hepta-bromo-DE (2009) A*
Hexabromobiphenyl (2009) A
PFOS, its salts and PFOSF (2009) B*
Hexabromocyclododecane (2013) A*
Deca-bromodiphenyl ether (2017) A*
Short-chain chlorinated paraffins (2017) A*
PFOA and its salt and PFOA-related compounds (2019) A*
Polychlorinated biphenyls (2001) A・C
Polychlorinated naphthalenes (2015) A*・C
Hexachlorobutadiene (2015) A・(2017) C

Polychlorodibenzo-*p*-dioxin (PCDDs), Polychlorodibenzofuran (PCDFs) (2001) C
Unintentional Production (Burning products, by-products)

Fig. 8.1 Chemicals listed as POPs under the Stockholm Convention on POPs, their use, regulatory categories, and listed year (as of January 2021). *DDT* 1-chloro-4-[2,2,2-trichloro-1-(4-chlorophenyl)ethyl]benzene, *DE* diphenyl ether, *PFOS* Perfluorooctane sulfonic acid, *PFOSF* Perfluorooctane sulfonyl fluoride

8.4 POPs Regulation in Japan before and after Stockholm Convention

8.4.1 Polychlorinated Biphenyls (PCBs) Waste

Polychlorinated biphenyls (PCBs), one of the most infamous chemicals among the POPs, were the causing agents of the Kanemi rice oil disease (Yusho) incident in 1968 in southern part of Japan. In the wake of this incident, the Japanese government made the chemical regulating law named "Law Concerning the Examination and Regulation of Manufacture, etc., of Chemical Substances (Chemical Substances Control Law)", one of the most advanced chemical regulating law at that time in the world and manufacture and import of PCBs were prohibited and PCB oil wastes were ordered to be stored. However, for more than 30 years, no PCB waste treatment facilities were constructed due to the oppositions by nearby residents. During the period, missing or illegal disposal of PCB waste often occurred.

Responding to the Stockholm Convention, the Japanese government made the "Law Concerning Special Measures for Promotion of Proper Treatment of PCB

Waste (PCB Special Measures Law)" in 2001 and Japan Environmental Storage and Safety Cooperation (JESCO) constructed five PCB waste destruction facilities over Japan under the national policy and subsidies [5]. As the opposition against incineration of PCB waste was very strong, chemical destruction was adopted as treatment method (plasma melting was added later). The deadlines of the treatment are set between 2018 and 2023 depending on the facilities. Coming closer to the deadlines, exploration of hidden PCB waste is now under way. The JESCO treatment is targeted at high concentration PCB waste (>5000 ppm). As large amount of low concentration PCB insulating oil waste was found afterwards, that cannot be handled by JESCO facilities due to their limited capacity, private hazardous waste treatment facilities were allowed to incinerate or to decontaminate below 5000 ppm PCB waste under the license given by the Ministry of the Environment, Japan (MOE) from 2010 [6]. Furthermore, demonstration tests in the direction of allowing incineration of up to 10% (100,000 ppm) PCB waste started in 2019 [7] and first license was given in 2020. Although it is a hindsight, it is questionable that construction of new treatment facilities with high-cost chemical destruction methods was necessary. Existing incineration technology must have been enough to cope with the PCBs waste if the government could convince the residents of its safety and merits at the starting stage.

8.4.2 *Organochlorine Pesticides*

Nine out of the 12 POPs listed at the start of Stockholm Convention were organochlorine pesticides (Fig. 8.1). All of them had been already banned or had not been used intentionally in Japan at that time. Likely, the 8 pesticides listed afterwards had been banned or their registrations had been expired at the time of their listing and no further regulations were necessary. However, there is one problem left unsolved. As banning the use of those pesticides, their stock was buried under soil in some remote forests. These may pose risk to groundwater in the years. Thus, monitoring, recovery, and disposal of the buried pesticides are underway.

8.4.3 *Industrial Chemicals*

Only one industrial chemical was listed at the start of Stockholm Convention, namely PCBs, which I have already elaborated above. From 2009, more industrial chemicals have come to be listed as POPs (Fig. 8.1). The Japanese government has made it a rule to designate them as the class I specified chemical substances (prohibition of production, import, and use in principle) under the Chemical Substances Control Law. However, collection of the products containing those chemicals under use has not been conducted based on the results of risk assessments (note: collection and destruction of PCB and dioxin containing pesticides has been underway). Those actions on new POPs have been taken only after their listing in Stockholm Convention and can be regarded as follow-up actions after international regulation. The examples are brominated flame retardants (BFRs: polybromodiphenylethers (PBDEs) and hexabromocyclododecane (HBCD)) and perfluorinated alkyl substances (PFASs: perfluorooctane sulfonic acid (PFOS), its salts and perfluorooctane sulfonyl fluoride (PFOSF); perfluorooctanoic acid (PFOA), its salts and PFOA-related compounds), and organochlorines (short chain chlorinated paraffins (SCCPs) and hexachlorobutadiene). BFRs, PFAS, and SCCPs were used in various industrial products distributed to consumers and in some industrial applications. To quickly cope with those, the Japanese government had to introduce new regulations and relevant industrial sectors had to develop alternative chemicals and/or processes.

It is highly desirable that screening and measures against potential POPs should have been taken before international listing based on the existing chemical re-inspection based on environmental risk assessment under the Chemical Substance Control Law. In addition, elimination of one (or a group of) chemical would normally results in the replacement with one or more of alternative(s) and comparative assessment in benefit as well as risk between the chemicals becomes necessary. In the case of high performance chemicals, performance of their alternative chemicals is often inferior to that of its predecessors, resulting in higher amount of their use. In other cases, plural alternative chemicals would be necessary for different applications and result in mixture pollution. In conclusion, a broad comparative assessment is necessary when introducing a regulation, taking expected future changes into accounts. It should be also noted that the science of mixture health effect is still its infancy.

8.4.4 Unintentional Products

Polychlorodibenzo-*p*-dioxins and polychlorodibenzofurans (PCDDs and PCDFs, hereafter referred as dioxins) can be formed anywhere with carbon and chlorine atoms and high temperature (such as in incinerators or thermal processes) and in chemical syntheses of chlorinated aromatic compounds. Those unintentionally formed chemicals are called as unintentional products. Dioxins, PCBs, polychlorinated naphthalenes (PCNs), hexachlorobenzene (HCB), pentachlorobenzene (PeCB), and hexachlorobutadiene (HCBD) are listed as unintentional POPs products in Stockholm convention (Fig. 8.1). Apart from industrial chemicals, unintentionally formed POPs cannot be eliminated just by banning their production.

In Japan, environmental pollution by dioxins came to be known in the 1980s and became a large social problem in the 1990s. To cope with the people's growing anxiety, the Japanese government established the "Law Concerning Special Measures against Dioxins" in 1999 and emission reduction measures were taken with strict emission standards on incinerators, thermal and chemical processes.

Under the law, target facilities are ordered to report their dioxin concentrations in their exhaust gas, wastes, and wastewaters every year and the MOE updates the national dioxin emission inventory annually [8]. The total emission decreased from around 8000 g-TEQ/y in 1997 to around 110 g-TEQ/y in the late 2010s (Fig. 8.2). Whereas the emission from waste incineration sector was reduced by 1/100 during the period, that from industrial sectors (mainly steel and non-ferrous metal industries) was reduced by around 1/10.

Concerning the other five unintentionally formed POPs, namely, PCBs, PCNs, HCB, PeCB, and HCBD, which had also been used as industrial chemicals, the MOE started to survey the emission sources and amounts of their emissions after their listing as POPs in the convention, in addition to the ban of their production, import, and use. As all the five can be formed in combustion and thermal process by-products, the emission reduction measures against dioxin were expected to reduce those emissions as well. The trends of amounts of PCBs and HCB emission were estimated and those of PCN, PeCB, and HCBD are underway [9]. The results indicated little and small reductions in emissions for PCBs and HCB, respectively, between 2002 and 2018 (Figs. 8.3 and 8.4). This meant that dioxin emission control measures were not so effective to the reduction of PCBs and HCB emissions. There is no regulation on the emission of unintentional POPs other than dioxins and PCBs and it is expected that publicized inventories will push facilities to take reduction measures. Cement combustion furnaces and metallurgical industry contributed much to the emission of PCBs and metallurgical industry and waste incinerators contributed much to that of HCB. The reliabilities of PCB and HCB emission inventories are still low compared with that of dioxins. Whereas the emissions of dioxins

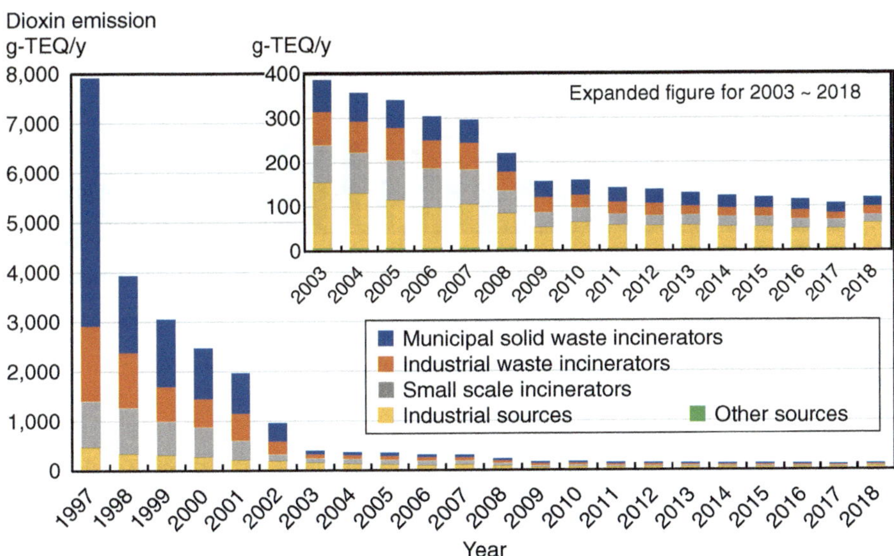

Fig. 8.2 Trends of total dioxin emission in Japan [8]

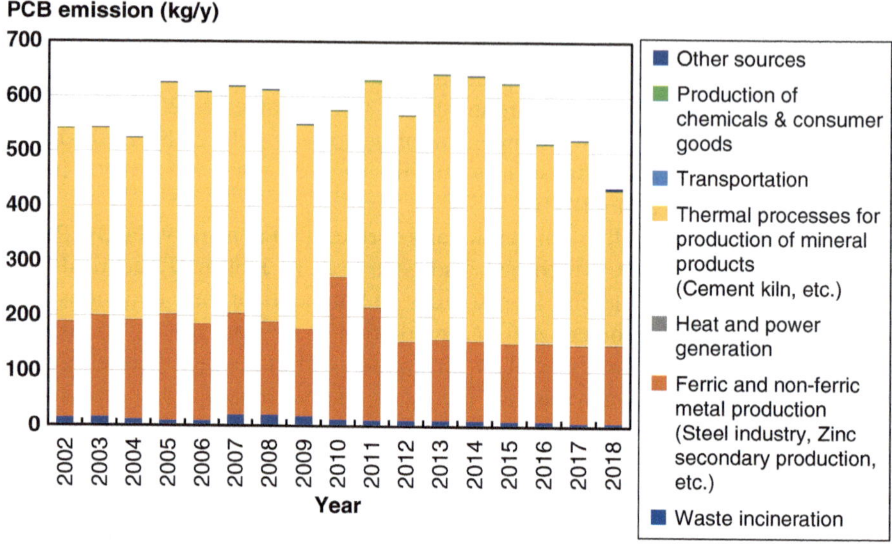

Fig. 8.3 Trends of polychlorinated biphenyls emission in Japan [9]

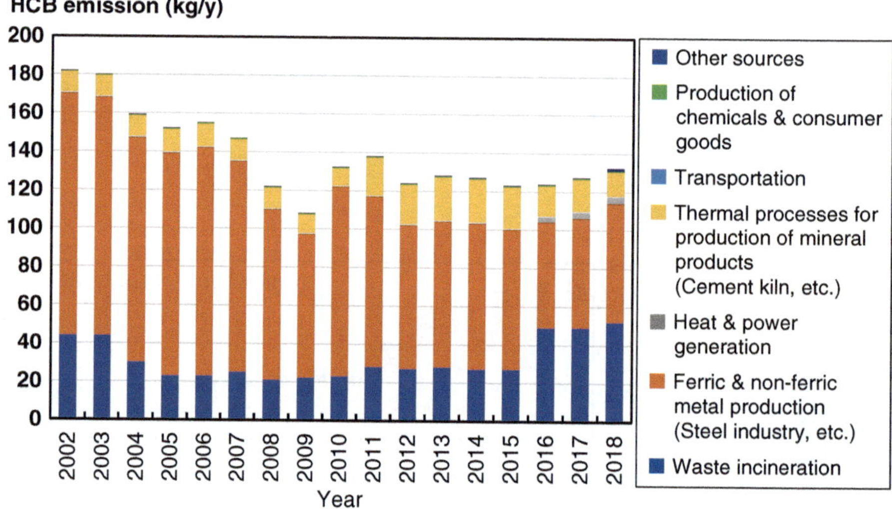

Fig. 8.4 Trends of hexachlorobenzene emission in Japan [9]

are to be reported by individual facilities every year under the special law, monitoring of other unintentionally formed POPs was carried out by the MOE for very limited facilities. There should be a better allocation of money and efforts for monitoring between dioxins and other unintentional POPs as large reduction of dioxin emission has been achieved.

8.5 Achievements in Dioxins and PCBs Management

Let us see the effect of POPs reduction measures based on the environmental and human monitoring by the MOE of Japan.

As for dioxins, discontinuation of wide spray of agrochemicals with dioxins as by-product in rice paddy field during the 1960s to early 1980s and the use of PCB containing products during the 1960s to early 1970s, as well as the regulation on dioxin emission sources since 1999 and the destruction of PCBs wastes since 2004, the level of dioxins (PCDDs, PCDFs, and coplanar PCBs) in human blood has decreased from the 20 pg-TEQ/g-fat in the 2000s to 10 pg-TEQ/g-fat in the 2010s (Fig. 8.5) [10]. Concerning the effect of PCBs waste destruction, Hirai & Sakai [11] estimated that PCBs evaporation from PCBs waste storage decreased from about 2700 kg/y in 2003 to about 1000 kg/y in 2013. They also regarded that reduction of PCB concentration in air (summer: 260 → 100 and winter: 110 → 57 pg/m^3) might be due to the progress of destruction and improvement in storage. This is a good example showing that regulation on sources and better management of waste stockpile have brought down the environmental contamination levels.

8.6 Challenges in Chemical Management in Japan

Japanese management of chemicals in products is still to be improved. As stated above, number of chemicals that had been widely used in products came to be listed as POPs. This means that premarketing assessment of new chemicals and re-inspection on existing ones under the Chemical Substances Control Law have not

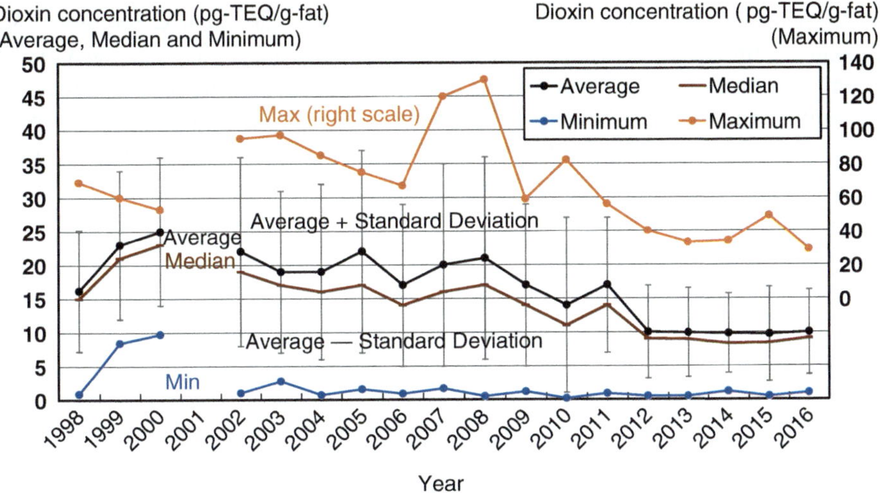

Fig. 8.5 Trends of dioxin concentration in Japanese human blood [10]

worked preventatively. This is partly because the definition of POPs under Stockholm Convention and that of class I or II specified chemical substances under the Chemical Substances Control Law are not the same. If the Japanese government, however, will continue to regulate POPs under the law, the two definitions should be harmonized so that the assessment system becomes capable of screening potential POPs. Otherwise, costly recall, recovery, and destruction of once circulated chemicals and products would be necessary.

In Japan, there is no basic law that covers all kinds of chemicals. The Chemical Substances Control Law is for general chemicals, however, some chemicals for specific uses, for example, pharmaceuticals, cosmetics, food additives, agrochemicals, fertilizers, etc., are excluded and covered by their specific laws. Those specific laws generally do not consider the risk through environmental exposure. This gap should be filled and other new hazardous effects such as occurrence of resistant bacteria and pest in the environment for antibiotics, antibacterial agents, or pesticides should be taken into regulatory systems.

Aside from government regulation, we must reexamine our daily life with respect to chemical usage. Health hazard incidents at home are often reported by improper use of chemical-containing products, for example, spray cans. While consumers pay close attention to food additives such as preservatives and food residues such as pesticides, they pay less attention to the side effects of household items such as insecticides, insect repellents, deodorants, air fresheners because of their convenience. Basic knowledge on the risks, appropriate handling, and disposal of chemical-containing products is essential.

As for the policy on unintentionally formed POPs, the MOE, Japan is making efforts in identifying sources and constructing emission inventory as shown in Figs. 8.2 and 8.3, though the budget for the efforts is rather limited. Those are providing novel and world leading results [12]. Plans for reduction measures will have to be constructed in the future.

With the increase of new substances and the discovery of new hazards (both human health and ecotoxicological risks), risk assessment and management of chemical substances is a never-ending task. In the future, it will become necessary to compare the risks and benefits between alternatives from a broad perspective for making the best choices. Furthermore, if Japan is to advocate environmentally advanced countries, we should constantly improve our policies to construct comprehensive chemical management system that can be a good example from the perspective of the Sustainable Development Goals (SDGs).

References

1. Agenda 21. United Nations Conference on Environment & Development Agenda 21. 1992. https://sustainabledevelopment.un.org/content/documents/Agenda21.pdf
2. Johannesburg Plan of Implementation. Division for Sustainable Development, UN. 2002. https://www.un.org/esa/sustdev/documents/WSSD_POI_PD/English/POIToc.htm. Accessed 15 Feb 2021.

3. SAICM. International conference on chemicals management. 2006. http://www.saicm.org/About/ICCM/tabid/5521/Default.aspx Accessed 15 Feb 2021.
4. UNEP. Global Chemicals Outlook II: from legacies to innovative solutions – implementing the 2030 Agenda for Sustainable Development. 2019. https://www.unep.org/resources/report/global-chemicals-outlook-ii-legacies-innovative-solutions. Accessed 15 Feb 2021.
5. JESCO. PCB waste treatment. 2001. https://www.jesconet.co.jp/eg/pcb/index.html. Accessed 15 Feb 2021.
6. MOE. Demonstration tests of PCB waste incineration (*in Japanese*). 2009. https://www.env.go.jp/press/106801.html. Accessed 15 Feb 2021.
7. MOE. Revision of relevant laws and regulations concerning the expansion of PCB waste subjected to the certified detoxification treatment facilities (*in Japanese*). 2019. https://www.env.go.jp/press/107555.html. Accessed 15 Feb 2021.
8. MOE. Dioxin emission inventory (*in Japanese*). 2020. https://www.env.go.jp/press/107882.html. Accessed 15 Feb 2021.
9. Government of Japan. The National Implementation Plan of Japan under the Stockholm Convention on Persistent Organic Pollutants, Modified in November 2020 (Received by the POPs Convention Secretariat on Dec. 12, 2020). 2020. http://chm.pops.int/Implementation/NIPs/NIPTransmission/tabid/253/Default.aspx. Accessed 15 Feb 2021.
10. MOE. The exposure to chemical compounds in the Japanese People – Survey of the exposure to chemical compound in human (2011). 2017. https://www.env.go.jp/chemi/dioxin/pamph/cd/2017en_full.pdf
11. Hirai Y, Sakai S. Environmental behavior of PCBs at waste storage sites and effects of PCB waste elimination measures. Mater Cycles Waste Manage Res. 2017;28(2):143–8. https://www.jstage.jst.go.jp/article/mcwmr/28/2/28_143/_article/-char/en
12. UNEP. List of national implementation plans. 2021. http://chm.pops.int/Implementation/NIPs/NIPTransmission/tabid/253/Default.aspx. Accessed 15 Feb 2021.

Chapter 9
Health Risk Assessment and Management of Asbestos Exposure to the Public in Japan

Naomi Hisanaga, Kiyoshi Sakai, and Eiji Shibata

Abstract Asbestos disasters are one of the major environmental issues associated with the industrial use of natural resources globally. There are enormous numbers of reports on occupational exposure to asbestos; however, comparing with these, reports on its non-occupational exposure are few. Herein, we report asbestos disasters in Japan and discuss health risk assessment and management of asbestos exposure to the public, using two instances that we experienced. One is an asbestos leakage incident at a subway station in Nagoya city. It is an example of asbestos exposure to the public caused by the removal of existing asbestos-containing materials after the ban on asbestos use in Japan. The other is the exposure derived from serpentinite. Serpentinite is a rock that is available in many places in Japan, often contains asbestos, and is used as a construction material and an industrial raw material. It is an example of health risks to the public derived from naturally occurring asbestos. Inhabitants in the vicinity of serpentinite outcrops are the subjects of exposure. There were many difficulties in risk assessment and management in both cases. Enhanced efforts to reduce asbestos exposure to the public through collaboration among various stakeholders are required.

Keywords Asbestos · Risk assessment · Public · Non-occupational exposure
Removal · Subway station · Serpentinite

N. Hisanaga (✉)
Aichi University of Education, Kariya, Japan

K. Sakai
Nagoya City University of Medicine, Nagoya, Japan
e-mail: sa-ka-i08240620@ma.medias.ne.jp

E. Shibata
Yokkaichi Nursing and Medical Care University, Yokkaichi, Japan
e-mail: eshibata@y-nm.ac.jp

T. Nakajima et al. (eds.), *Overcoming Environmental Risks to Achieve Sustainable Development Goals*, Current Topics in Environmental Health and Preventive Medicine, https://doi.org/10.1007/978-981-16-6249-2_9

9.1 Asbestos Disasters in Japan

Our first involvement with asbestos was in 1982, when we investigated an automobile brake and clutch reprocessing plant [1]. Subsequently, in 1986, we started working on the prevention of asbestos diseases among construction workers [2], and have been working on it ever since. Reflecting back, we think that we underestimated the hazards caused by asbestos. For instance, we determined the concentration of airborne dust while cutting a board containing asbestos with a circular electric saw at construction sites without a mask (Fig. 9.1).

The current scenario is a repercussion of the negligence in undertaking control measures against asbestos-related diseases. Figure 9.2 shows the number of patients with mesothelioma and asbestos-related lung cancer recognized under the Industrial Accident Compensation Insurance Law and the Asbestos Health Damage Relief Law since 1990. In Japan, the social recognition of the health risk of asbestos has changed drastically before and after the Kubota Shock in 2005, when a mass outbreak of mesothelioma in people living near a former large asbestos cement pipe plant came into light [3]. It triggered the enactment of a new law, the Asbestos Health Damage Relief Law, in 2006. The subjects of the new law are workers unable to claim the industrial accident compensation due to the passing of the statute of limitations, and non-workers that refer to people such as housewives and self-employed people not covered by the Industrial Accident Compensation Insurance Law. Figure 9.2 illustrates the changes in the annual number of people compensated or relieved as mesothelioma or lung cancer caused by asbestos since 1990. The total number of people certified under the two laws between 1990 and 2019 was 31,823, consisting of 22,544 patients with mesothelioma and 9279 with lung cancer.

The Environmental Restoration and Conservation Agency of Japan has surveyed the asbestos exposure history of non-workers relieved by the Asbestos Health Damage Relief Law. The survey categorized exposure history into four groups: (a)

Fig. 9.1 Heavy dust exposure to a carpenter during cutting asbestos-containing board with an electric circular saw at a construction site in 1987. The airborne asbestos (type: amosite) concentration of 131×10^3 f/L at his respiratory zone, corresponds to 873 folds of the current administrative level (150 f/L) of asbestos in Japan

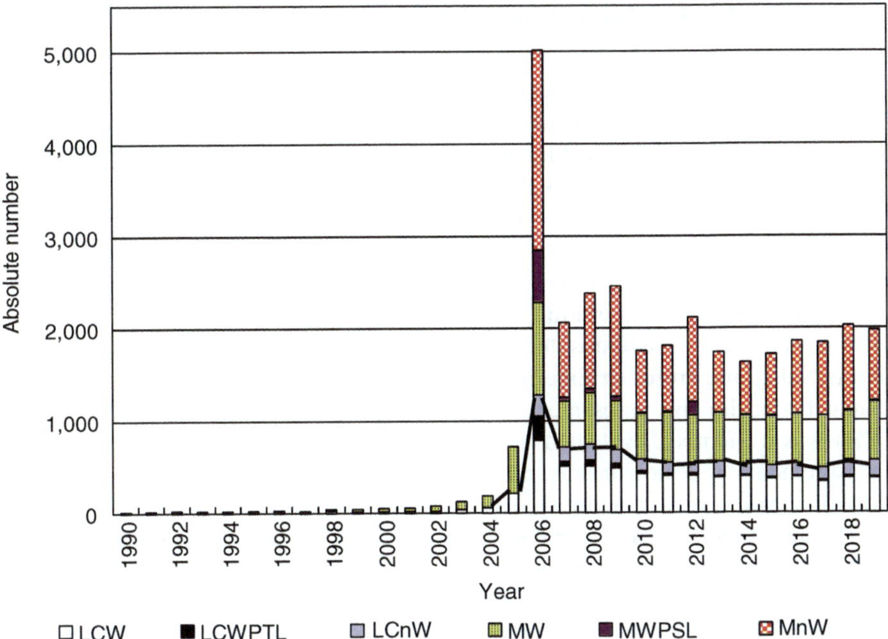

Fig. 9.2 Changes in the annual number of workers compensated by the Industrial Accident Compensation Insurance Law, and workers who have lost the right to claim compensation owing to passing the statute of limitation and non-workers relieved by the Asbestos Health Damage Relief Law, as mesothelioma or lung cancer due to asbestos in Japan. *LCW* Lung cancer worker, *LCWPTL* Lung cancer worker; Past the statute of limitation, *LCnW* Lung cancer non-worker, *MW* Mesothelioma worker, *MWPSL* Mesothelioma worker, Past the statute of limitation. *MnW* Mesothelioma non-worker. Source: The yearly summary on asbestos diseases compensation by the Ministry of Health Labor and Welfare, and the questionnaire survey report on asbestos exposure of non-workers relieved by the Asbestos Health Damage Relief Law from 2006 to 2018 by the Environmental Restoration and Conservation Agency of Japan

direct or indirect occupational exposure; (b) household exposure; (c) exposure during entry into asbestos handling facilities and living rooms or office space where asbestos was sprayed on walls and ceilings; and (d) presence of asbestos handling facilities in the vicinity of residence, school, workplace, or unidentified exposure. Among 9670 patients with mesothelioma and lung cancer who responded to the survey between 2006 and 2018, 57.9% were in group (a), 2.3% in group (b), 1.9% in group (c), and 37.9% in group (d) [4]. The fact that non-occupational exposure (b) to (d) exceeds by 40% indicates the damage caused to citizens at each stage of asbestos use to disposal. In this report, we have focused on asbestos exposure to citizens and discussed the incident at a subway station as an example of exposure to citizens during the removal of asbestos products, which occurs daily throughout Japan, and the problem of asbestos-containing serpentine rock as a typical example of unknown exposure to citizens from the viewpoint of risk management and sustainable development.

9.2 Asbestos Leakage at a Subway Station

On December 12, 2013, work began to remove sprayed-on asbestos (type: crocidolite) from the walls and ceiling of the ventilation machine room of a subway station in Nagoya city. A public health center in the city measured the concentration of airborne asbestos at the walkway for station users in front of the entrance door of the room and at a point on the ground beside the ventilation tower emitting air from the underground station to the outside. The next day, the results showed that these concentrations were 1000 f/L in the former and 4 f/L in the latter and the work was halted. The city's transportation bureau set up an "Investigation Committee on Health Measures for Asbestos dust leakage at Rokubancho Station" (chaired by Dr. Tamie Nakajima, Professor Emeritus at Nagoya University) and commissioned it to conduct a health risk assessment. The study group, of which one of the authors (NH) was a member, evaluated health risks based on on-site inspections, actual measurements of airflow inside and outside the station, and a map of asbestos dust diffusion inside the station drawn over time by simulation [5].

The investigation revealed that the contractor of the work did not follow the regulations for wetting to reduce dust emission during asbestos removal, the maintenance and use of the negative pressure dust removal equipment were inappropriate, the louver of the door between the removal site and the station user walkway was not sealed, and these problems were overlooked during the on-site inspection by the public health center on the day before the removal. Consequently, the working environment at the removal site became highly dusty, and simultaneously, a considerable amount of asbestos dust leaked into the walkway for station users (Fig. 9.3).

The health risks were assessed by calculating the lifetime excess cancer risk (mesothelioma and lung cancer) due to asbestos exposure using three methods: the U.S. Environmental Protection Agency [6], the World Health Organization [7], and the method proposed by Hughes of Tulane University [8]. Among the three methods, using the Hughes' method, which was the highest risk estimate, the risk per 100,000 people was 0.022, 0.011, and 0.474 for infants, adults, and station workers, respectively. These figures are less than the lifetime risk of one person per 100,000 that the Central Environment Council of Japan concluded to be a practically safe level for exposure to air pollutants with no toxicity threshold in 1996. Based on these results, Nagoya city decided not to compile a list of station users on the day of the accident or to conduct a special long-term medical checkup, but to establish a health counseling system for station users and station workers and to take measures to prevent the recurrence of accidents [5].

Reflecting back on this case, there are several points that we would like to describe. First, immense efforts and a considerable amount of money were needed to deal with this case; asbestos leakage could have been prevented if the ventilation louver of the door had been sealed. This highlights the importance of appropriate risk management. Second, although there was no way other than to do so, the amount of asbestos exposure had to be estimated using insufficient data measured by the public health center, and health risks were estimated using methods based on

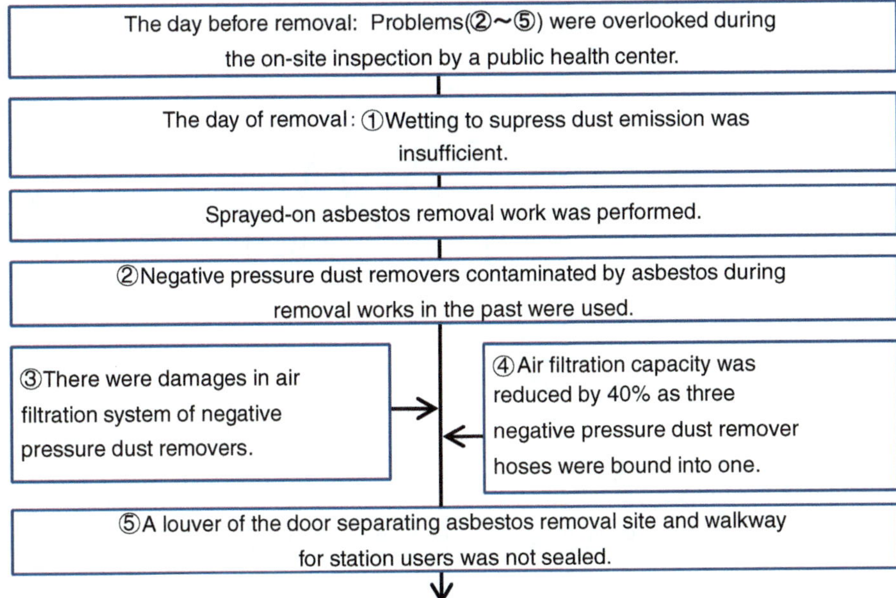

Fig. 9.3 Factors causing the asbestos leakage during sprayed-on asbestos removal work at a subway station in Nagoya city

the epidemiological data gained from years of exposure in the workplaces. There has been little evidence justifying the extrapolation of long-term epidemiological findings to short-term exposure in asbestos health risk assessment. Third, the reaction of Nagoya city in dealing with this asbestos leak case was exceptional. Although asbestos exposure to citizens during building renovation and demolition occurs frequently in Japan, no appropriate subsequent measure has been taken in many cases. Fourth, the investigation committee meeting and the minutes were open to the public; however, the city's transportation bureau received only a few public responses, which suggests a low public concern for this incident.

9.3 Asbestos Exposure in Areas with Serpentinite

Serpentinite is widely distributed in Japan and has a massive, foliate, and muddy form. It is used as crushed stone, stone material, a raw material for phosphorus fertilizer, and steel mill fluxing material [9]. The annual trend in the amount of serpentinite mined in Japan reported by the Agency for Natural Resources and Energy of Japan is shown in Fig. 9.4, along with the amount of asbestos imported and domestically mined. The amount of serpentinite mined was much larger than the amount of

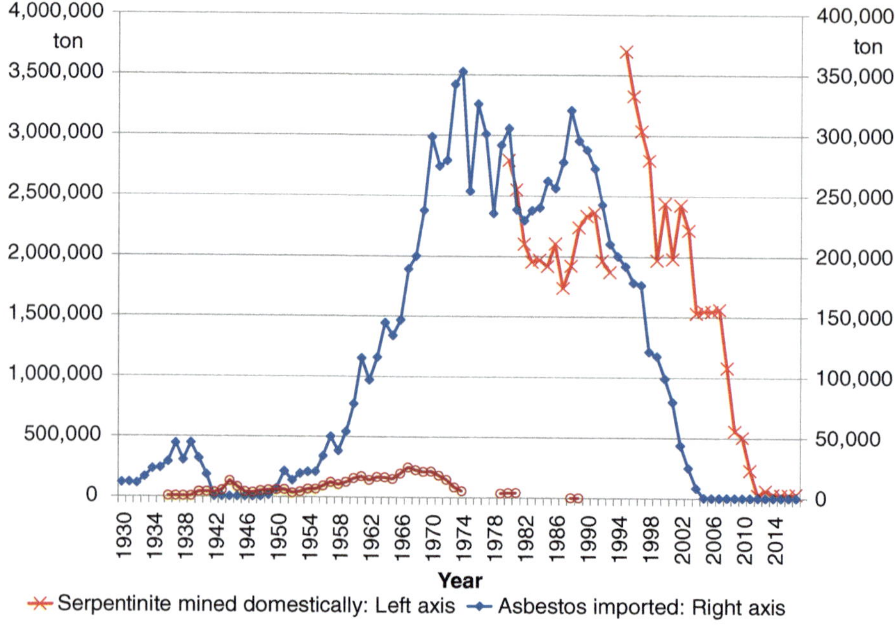

Fig. 9.4 Changes in the annual amounts of asbestos imported and mined domestically, and serpentinite mined domestically in Japan. Source: Trade statistics of Japan by the Ministry of Finance, yearbook of current production statistics on ceramics and building materials by the Ministry of Economy, Trade and Industry, and report on business condition of quarrying industry by the Agency for Natural Resources and Energy of Japan

asbestos imported and mined domestically. So far the data we obtained, the largest amount of serpentinite mined was 3.96 million tons in 1995. This amount has declined sharply in recent years; however, it was still 9200 tons in 2019. We analyzed serpentinite rocks from seven sites, namely, Hokkaido, Saitama, Chiba, Kanagawa, Aichi, Mie, and Kumamoto prefectures, and detected asbestos in all the specimens, including seven sites with chrysotile and four sites with tremolite/actinolite [10]. In a serpentinite outcrop area in eastern Aichi Prefecture, tremolite/actinolite was also found in the soil in the agricultural fields, with higher concentrations in the air than in the control area [11] and higher concentrations in the lung tissues of residents than in the reference area [12]. There was also a man who grew up in the same area and died of mesothelioma at 35 years of age without any occupational history of asbestos exposure, who had high levels of chrysotile and actinolite in his tumor tissue [10]. Asbestos exposure to citizens can occur from living (Fig. 9.5), civil engineering, agricultural work, and stone processing in areas of serpentinite distribution. Even now, not a small amount of serpentinite rock has been mined; however, there are many unknowns about where and how it is being used and how citizens are exposed to it. Currently, in Japan, the production and use of materials containing more than 0.1% of asbestos is prohibited; however, there is a great

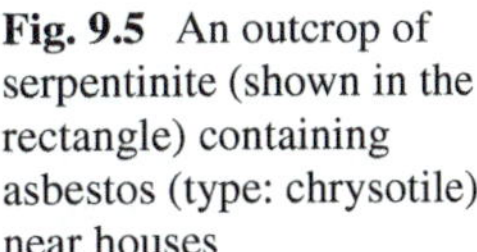

Fig. 9.5 An outcrop of serpentinite (shown in the rectangle) containing asbestos (type: chrysotile) near houses

possibility that serpentinite is still available in the market, which violates the law. Many people live in areas where serpentinite is distributed without knowing that it contains asbestos. As countermeasures against asbestos exposure derived from serpentinite are related to land use, development, and land prices, they must be performed with the consent of many stakeholders.

9.4 Health Risk Management and Sustainable Development

Asbestos was used in large quantities for many purposes, and even after its hazard was discovered, countermeasures were delayed, and the damage occurred not only to the manufacturers and users but also to citizens. The risk of exposure to asbestos, which does not readily decompose, existed at almost every stage of the life cycle of the products and buildings where it was used. Even after the ban on new production, asbestos exposure from decayed products, product removal, and serpentinite continues. Therefore, the number of people subject to health risk management is enormous. In contrast, even for workers, it is difficult to accurately determine the amount of asbestos exposure in the past, and the quantitative relationship between exposure and health effects is still unclear; therefore, complete health risk management is not easy. This is even worse for general citizens. Issues due to asbestos-related diseases still persist, and the prevention of these diseases should be tackled, while the simultaneous application of the lessons learned from the asbestos disaster for future sustenance. The lessons learned from the asbestos disaster include delays in dust control measures, delays in regulating the use of asbestos, and the mass outbreak of asbestos-related malignancies among citizens living near factories. Together with these lessons, a lack of experts and weak cooperation among related academic fields that were necessary to solve these problems existed in the background of asbestos disasters. Additionally, there are many other detrimental factors in the living and working environments, and there is the possibility of large-scale damage occurring due to these factors in the future. For sustainable development, citizens' ability to cope with various harmful factors must be enhanced. Based on our experiences as faculties in universities and public institutes, we would like to observe progress in this area, including the enhancement of school education.

References

1. Sakai K, Hisanaga N, Shibata E, Ono Y, Takeuchi Y. Asbestos exposure during reprocessing of automobile brakes and clutches. Int J Occup Environ Health. 2006;12:95–105.
2. Hisanaga N, Hosokawa M, Sakai K, Shibata E, Huang J, Takeuchi Y et al. Asbestos exposure among construction workers. Proceedings of 7th international pneumoconiosis conference. Pittsburg. 1991. p. 1053–1058.
3. Kurumatani N, Kumagai S. Mapping the risk of mesothelioma due to neighborhood asbestos exposure. Am J Respir Crit Care Med. 2008;178:624–9.
4. Agency for Natural Resources and Energy of Japan. Questionnaire survey report on asbestos exposure history of people recognized to be asbestos-related diseases from 2006 to 2018 under the asbestos health damage relief system. 2020. p. 56–60.
5. Investigation Committee on Health Measures for Asbestos dust leakage at Rokubancho Station. Report of the Investigation Committee on Health Measures for Asbestos dust leakage at Rokubancho Station. Nagoya City Transportation Bureau; 2017. https://www.kotsu.city.nagoya.jp/jp/pc/ABOUT/TRP0002145.htm
6. U.S. Environmental Protection Agency. Asbestos; CASRN 1332-21-4. 1988. p. 1–14.
7. WHO Regional office for Europe. Asbestos. In: Air quality guidelines for Europe. 2nd ed. Copenhagen: WHO Regional office for Europe; 2000. p. 128–35.
8. Hughes JM, Weill H. Asbestos exposure-quantitative assessment of risk. Am Rev Resp Dis. 1986;133:5–13.
9. Yoshida K. Serpentinite. In: Knowledge on mineral products and business. Tokyo: Tsusho-Sangyo-Chosakai; 1992. p. 482–7.
10. Hisanaga N, Sakai K, Natori Y. Asbestos exposure risk derived from naturally occurring asbestos-containing rocks and soils and its health effects in Japan. Proceedings of 4th Meeting on comprehensive measures against asbestos issue. Tokyo. 2016. p. 22.
11. Sakai K, Hisanaga N, Kohyama N, Shibata E, Takeuchi Y. Airborne fiber concentration and size distribution of mineral fibers in area with serpentinite outcrops in Aichi prefecture, Japan. Ind Health. 2001;39:132–40.
12. Sakai K, Hisanaga N, Okuno M, Kohyama N, Shinohara Y, Shibata E, et al. Concentration and fiber size of asbestos in lungs of residents living close to the serpentinite area. Jpn J Public Health. 1996;43:551–62.

Chapter 10
Marine Plastic Pollution: Chemical Aspects and Possible Solutions

Hideshige Takada, Misaki Koro, and Charita S. Kwan

Abstract Plastics are formed from C–C single bonds which are flexible making them easily transformable. Thus, plastics have been widely utilized by the global society. However, the C–C single bond is easily breakable by UV radiation. Eventually, plastics are fragmented into microplastics (<5 mm) and ultimately will be washed off to marine environments. Microplastics are ubiquitously dispersed in marine environments, i.e., surface water, beaches, and bottom sediments. Plastics and microplastics are taken up and ingested by various marine organisms ranging from zooplanktons to whales. Concentrations of microplastics in urban coastal waters are considered to be critical in terms of particle toxicity. In marine environments, plastics retain hydrophobic additives and concentrate persistent organic pollutants (POPs) from surrounding waters. Recent field observations and exposure experiments demonstrated that microplastics accelerate indirect exposure of endocrine-disrupting additives to humans, through fragmentation of plastics, their ingestion by marine organisms, leaching of the additives to digestive fluids and their bioaccumulation and trophic transfer. The additives have the potential to disrupt endocrine and immune systems of humans. As a precautionary action, usage and consumption of plastics in our society should be reduced to protect human health. Reduction of plastic consumption is also deduced from the view of mitigating the impacts of global warming, resource efficiency, and sound waste management.

Keywords Microplastics · Additives · Endocrine-Disrupting Chemicals (EDCs) Persistent Organic Pollutants (POPs) · Sediment · Anthropocene

H. Takada (✉) · M. Koro
Tokyo University of Agriculture and Technology, Tokyo, Japan
e-mail: shige@cc.tuat.ac.jp

C. S. Kwan
University of the Philippines Diliman, Quezon City, The Philippines
e-mail: cskwan@up.edu.ph

T. Nakajima et al. (eds.), *Overcoming Environmental Risks to Achieve Sustainable Development Goals*, Current Topics in Environmental Health and Preventive Medicine, https://doi.org/10.1007/978-981-16-6249-2_10

10.1 Microplastics: Here, There, and Everywhere

On a global basis, 400 million tons of plastics are produced annually [1]. They consume ~8% of global petroleum production. Half of the produced plastics are single-use plastics, e.g., plastic bottles, food containers, plastic bags, and other packing materials. Some of them escape from waste management operations and are carried into rivers through surface runoff, and finally into the oceans. The volume of plastics entering into the oceans is estimated at several million tons per year (i.e., 5–13 million tons/y, [2]). When floating on the sea surface or stranding on beaches, they are exposed to UV radiation and get fragmented into smaller plastics, then into microplastics (i.e., plastics with less than 5 mm in diameter). Microplastics (MPs) are also generated in terrestrial environments. Plastics trashed on land surfaces and plastic products used outdoors, e.g., transport vehicle tires or artificial lawns are fragmented and weathered by physical impacts and UV radiation. Most of these MPs as well as large plastic trashes will be finally washed off to the oceans by surface runoff such as street runoff [3]. There are several other origins of MPs in marine environments, i.e., resin pellets, microbeads in personal care products, chemical fibers from synthetic textiles, shavings from sponge of synthetic fibers, and many others.

10.2 Anthropocene: Ubiquitous Occurrence of Microplastics

As discussed above, plastics from different polymers in a wide range of sizes are supplied to the oceans from various sources. During transport into the oceans, they behave differently depending on their sizes and densities. Lighter polymers such as polyethylene (PE) and polypropylene (PP) float on liquid surfaces. However, floating plastics larger than 5 mm can be transported inshore by wind and are stranded on beaches [4], where they will be fragmented into MPs through UV radiation and heat. MPs are then washed offshore and will be eligible to long travel, e.g., to thousands of kilometers, even to remote islands; thus, dispersing plastic pollution all over the world. They are also accumulated in certain parts of the ocean, e.g., in gyres due to circulating ocean currents and large wind movements [5]. Consequently, huge amounts of plastics (i.e., 5 trillion pieces or 270,000 tons of plastics, [6]) are floating in our oceans.

On the other hand, polymers heavier than water such as polyvinyl chloride (PVC) and polyethylene terephthalate (PET) are accumulated in coastal sediments. Even lighter polymers such as PE and PP settle to the bottom when they become smaller than 1 mm through biofilm formation which makes the plastic particles heavier. Lighter polymers can also be ingested by small organisms, and subsequently excreted as fecal pellets or incorporated into the corpse. These are accumulated in bottom sediments through settling. The amounts of MPs in bottom sediments are greater than those in surface waters. Bottom sediments can act as a huge sink for

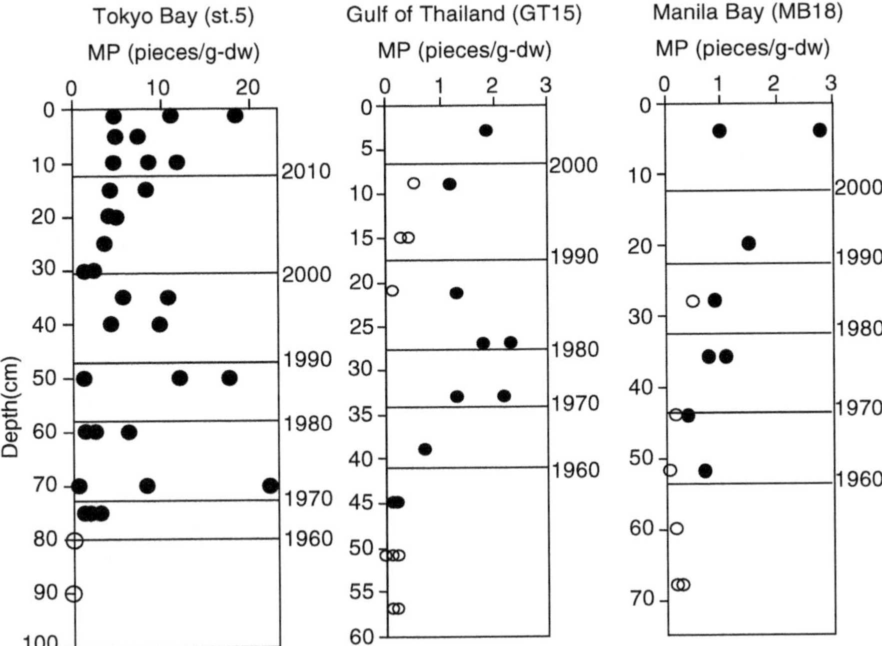

Fig. 10.1 Vertical profiles of microplastics in sediment cores from Tokyo Bay, Gulf of Thailand, and Manila Bay. Estimated sedimentation age is indicated on light Y-axis of each panel. Solid symbols: significant detection; open symbols: insignificant values (below corresponding procedural blanks). Details including sample collection and treatment are available in Koro et al. [8]

MPs [7]. In the sediment cores obtained from Asian countries, a large number of MPs were accumulated in layers corresponding to the 1960s and beyond (Fig.10.1). The results are in accordance with the Anthropocene Epoch wherein human activities significantly impact the Earth's geology, atmosphere, and global material cycles [9].

10.3 Plastic Ingestion: Threat to Biodiversity and Food Security

Plastic and plastic fragments in the oceans are ingested by various marine organisms depending on their sizes. Ingestion of larger items (e.g., ~ cm) by large marine organisms such as whales, sea turtles, and seabirds has been reported since the 1970s [10]. As of 2016, 165 species of seabirds, corresponding to half of the total species of seabirds around the globe have been reported to ingest plastics [11]. The ingestion rate by marine organisms has been increasing [12]. Recently, ingestion of MPs by smaller organisms, i.e., shellfish, fish, crustaceans, and zooplankton, was reported. As of 2020, 427 species of fish have been reported to ingest MPs [13]. Approximately 80% of anchovies caught from Tokyo Bay, Japan contained ~1 mm microplastics in their digestive tracts [14]. Smaller MPs down to 0.01 mm were detected in bivalves, crab, and benthic fishes in Tokyo Bay.

As larger plastics pose physical impacts on larger marine animals such as seabirds and sea turtles [15], MPs also present a physical threat to smaller organisms such as fish and bivalves. Laboratory experiments demonstrated that MPs induce particle toxicity such as reduced body growth, physiological stress, and hormonal dysregulation [16]. According to the results from conventional monitoring which utilized a Neuston net with 300 μm mesh in collecting the MPs, concentrations of MPs (>300 μm; e.g., ~ 0.01 particles/L for Tokyo Bay) [17]) were much lower than the threshold concentration (6.65 particles/L [18]), below which adverse effects of particle toxicity are not likely to occur. However, a higher number of MPs were observed when a smaller range of MPs (down to 10 μm) was measured. In an estuary of Tokyo Bay, MP concentrations (>10 μm) ranged from 1 to 4 particles/L [3] which is close to the threshold for particle toxicity. The current status of MP pollution can be considered critical in terms of particle toxicity. Measurement of smaller MPs (<300 μm) in marine environments should be conducted to assess the impacts of MPs on marine organisms. Furthermore, chemical toxicity associated with MPs should be examined.

10.4 Warning Bell: Chemical Threat from Marine Plastics and Microplastics

Due to the hydrophobic nature of plastics, MPs absorb and accumulate persistent organic pollutants (POPs) such as polychlorinated biphenyls (PCBs) and organochlorine pesticides (DDTs and HCHs) from surrounding seawater. Although regulated by the Stockholm Convention, POPs are present in coastal waters as legacy pollutants. Most POPs are hydrophobic; therefore, they are accumulated in MPs due to hydrophobic interaction, as demonstrated by the International Pellet Watch (http://www.pelletwatch.org/).

Most plastic products contain additives such as plasticizers, UV stabilizers, antioxidants and flame retardants to maintain and/or improve their properties. On a per weight basis, additives represent ~7% of the production of plastics [1]. Some of them are classified as endocrine-disrupting chemicals (EDCs) such as bis(2-ethylhexyl)phthalate, benzotriazoles (UV-P, UV-329), nonylphenol, bisphenol A (BPA), and polybrominated diphenyl ethers (PBDEs) [19]. As additives are basically hydrophobic, they have a high affinity with plastics. They are compounded into the plastic matrix to avoid being easily leached out during normal usage. The hydrophobic additives are retained in marine plastics [20], including MPs [21]. However, once they are ingested by marine organisms, the hydrophobic additives are released from the plastic into the digestive fluid due to the lipophilic and/or surface-active components of the digestive fluid [22], and the small sizes (i.e., larger specific surface area) of the MPs [23]. The released additives are incorporated into the internal biological system and will be accumulated in the biological tissues [12] (Fig. 10.2). The transfer of hydrophobic additives from ingested plastics to the tissues and their bioaccumulation has been evidenced by field observations in seabirds [24] and in crustaceans [25], and by exposure experiments using seabirds [26], crustaceans [25], and fish. Subsequent trophic transfer in the food web carries the hydrophobic additives to humans. MPs should be considered as the vehicle of indirect exposure to endocrine-disrupting additives in humans (Fig. 10.2).

The number of field observations to prove the adverse effects of plastic-mediated chemical exposure is still limited. A study by Lavers et al. [27] focusing on flesh-footed shearwaters in Australia, reported changes in the blood chemistry of the seabirds. In individuals with more plastic ingestion, the cholesterol concentrations in the blood were higher than those with less plastic ingestion. Also, a decrease in calcium concentration in the blood was observed in seabirds with more plastic ingestion. A decrease in blood calcium may cause a thinner eggshell leading to a decrease in bird population. This is a phenomenon similar to the disruption of calcium transport in the body induced by DDT pesticide in the 1960s, as described in "Silent Spring" by Rachel Carson [28]. Plastics contain chemicals that disrupt the internal system of biota, e.g., the reported abnormality in blood chemistry. More abnormalities may occur if plastic pollution remains unabated. We have to consider

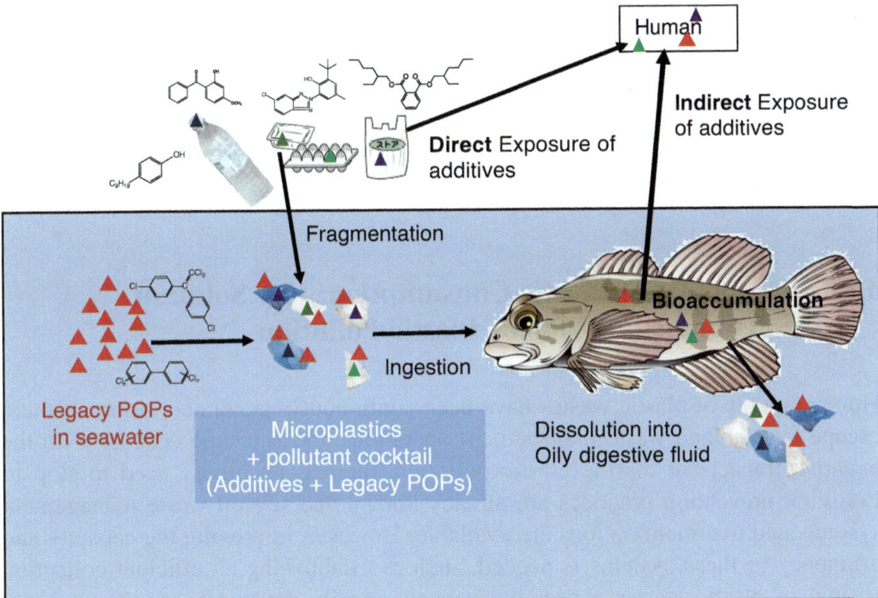

Fig. 10.2 Conceptual model of microplastic-mediated exposure of additives and persistent organic pollutants to humans

the abnormality in the seabirds as canary in the coal mine and take concrete actions to reduce plastic pollution.

Direct and indirect exposures of humans to EDCs have been increasing due to the increased usage of plastic products. The occurrence of MPs in the marine environment, their ingestion by marine organisms, and consumption of MP-contaminated seafood further accelerate the exposure of humans to EDCs. Plastic additives have been detected in the human body. The plastic additive BPA which is a compositional monomer of polycarbonate and epoxy resin was detected in women's blood [29]. More alarming is the significant detection of BPA in the blood of endometriotic women; whereas, it was not detected in the healthy group [29]. Furthermore, BPA together with PCBs and DDE were detected in human umbilical cord blood, suggesting exposure of the fetus to plastic additives through the mother [30]. The problem with risk assessment to exposure of plastic additives and their effects is that the resultant disorder will be seen many years later or even across generations. Abnormalities of the human reproductive system, e.g., an increase in the number of breast cancer cases and endometriosis have been observed. Decrease in sperm counts of European adults has been reported [31]. Because there are many other factors that affect the reproductive system, direct correlation of such abnormality to exposure to plastic additives is difficult to prove. However, it has been established that some plastic additives have the ability to decrease the reproductive ability of humans. Furthermore, some plastic additives have been documented to disrupt the immune system through biological processes. Exposure to endocrine-disrupting plastic additives can lead to diseases. It is stated that exposure to such chemicals is related to the COVID-19 pandemic [32]. Wu et al. (2020) have identified biological pathways as possible targets of EDCs and as contributors of the severity of COVID-19 [33]. Therefore, there is a need to quantitatively evaluate the contribution of plastic additives to endocrine disruption and immune depression. As a precautionary action, usage and consumption of plastics in our society should be reduced to protect human health.

10.5 Reduction of Plastic Consumption: Key Solution to Human and Marine Plastic Pollution

Huge amounts of plastic wastes have been continuously generated. Some of them escape from inland waste management operations and enter the oceans. With the negative impacts of marine plastic pollution, there is an urgent need to stop it. Pollution prevention practices are already known and several waste management systems and treatment options are available. However, improving the capacity and efficiency of these systems is needed, such as establishing an efficient collection system of plastic wastes especially in economically developing countries. Final waste treatment options are also crucial to ensure proper containment of plastic wastes. Landfilling of waste is a common waste management practice in many countries. However, we have limited areas for landfills. Furthermore, based on our studies on landfill sites [34, 35], landfill leachates discharge toxic plastic-derived chemicals (additives, monomers, and oligomers) to rivers and groundwater. Extremely high concentrations of BPA, nonylphenol, and PBDEs were detected in landfill leachates from Asian countries [36]. Furthermore, landfilling of plastics with biodegradable organic wastes such as kitchen wastes would increase the toxicity of additives such as PBDEs by anaerobic reactions which are caused by the degradation of organic waste [35]. Plastics should be segregated from organic wastes before dumping them in landfill sites. BPA in leachate from a closed landfill site is still contaminating a river running through Tokyo [37]. This is a case of legacy pollution, wherein waste plastics in landfill sites serve as a long-term source of toxic chemicals. Landfilling of plastics is not a sustainable option.

Incineration, waste-to-energy, or thermal recovery is another option. However, incineration of plastics produces toxic products such as dioxin and polycyclic aromatic compounds. We can prevent the emission of toxic chemicals by using sophisticated technology. However, this involves a huge cost to construct, operate and maintain modern incinerators throughout their lifetimes. For example, the construction of a modern incinerator with negligible dioxin emission to cover 400 thousand inhabitants costs USD100 million with an annual operating cost of USD2 million during its lifetime of 30 years. In addition, remnants of decommissioned incinerators contain extremely large amounts of hazardous materials. Thus, incinerators are a heavy financial and hazardous burden to future generations. It is not sustainable. Fundamentally, incineration of plastics causes net emission of CO_2 and is not consistent with the Paris Agreement as long as plastic is made of fossil fuel. From the viewpoint of the global carbon cycle, the combustion of fossil-fuel-based plastics is not circular.

As discussed above, landfilling and incineration are not clean and sustainable options for the treatment of plastic wastes. We need to minimize the amounts of plastic wastes to be landfilled or incinerated. Toward zero plastic waste, the key is the 3Rs of waste management (reduce, reuse, recycle). Among the 3Rs, reduction is the top priority. Recycling requires energy and cost. Breaking down of carbon-carbon single bond decreases quality of plastics; therefore, it is impossible to recycle plastics infinitely. Chemical recycling of plastic wastes through resynthesis to petroleum hydrocarbons to produce the same or higher quality plastics requires huge energy. Problems associated with waste-to-energy and chemical recycling are discussed in detail in Bell and Takada [38]. Thus, recycling is a limited option making reduction and reuse of plastics the top priority.

Replacement of petroleum-based plastics with biomass (i.e., paper and wood) and biomass-based plastic is an important and promising option, especially in tropical countries where biomass is abundant. However, biomass-based replacement should be taken with precautionary measures as this can be in conflict with food production and can also promote deforestation. If we replace all the petroleum-based plastics we use today with cellulose-based plastics, it may lead to severe deforestation and will accelerate global warming. The use of “biodegradable” plastic can be an option to mitigate marine plastic pollution. However, their degradability depends on environmental conditions, e.g., the difficulty to degrade under anaerobic conditions in bottom sediments. Microplastics made of polycaprolactone, a kind of biodegradable plastic, were detected in the bottom sediments of Tokyo Bay [7]. Biodegradable plastics should be treated in a closed degradation system such as a composter before discharging to the environment. Both biomass-based and biodegradable plastics will still require compounding with additives, and their safety should be carefully examined and verified. Thus, the remaining option is to reduce the current consumption rate of plastics products.

Based on the discussions above, the following specific options are recommended:

- Regulation on single-use plastics, especially PET bottles and other unnecessary plastic products, e.g., plastic straws, eating utensils, etc.
- Establishment of a social system of accountability: (a) increase public awareness on plastic pollution and the impacts of microplastics; (b) promote efficient garbage collection and segregation, and proper recycling of plastics; (c) adherence to international regulations, especially stopping the international trade of (plastic) wastes, "surplus goods" to developing countries.
- Development of products and packaging materials designed to facilitate reuse and safe recycling.
- Innovation of logistics to reduce carbon footprint and plastic consumption.
- Promotion of the utilization of biomass and biomass-based plastics together with the needed precautionary measures. and the proper treatment system.
- Careful evaluation of the toxicity of plastic additives and replacement with safer additives.

There is no single solution to address plastic pollution, but smart combinations of any of these options are necessary for individual countries to consider depending on their socio-economic conditions. At the core of these options, promotion of the 3Rs of waste management, especially reduction of plastics is the key. The COVID-19 pandemic increases the volume of plastic products consumed (personal protective equipment (PPE) of health workers, synthetic face masks and plastic face shields, more use of plastic packaging materials for delivered and take-out food, and other consumer products). However, as we described above, increased discharge of plastic wastes to marine environments would cause increased loads of EDCs to the marine ecosystem and eventually to humans, which would potentially decrease immunity of humans and would accelerate the severity of COVID-19, which in turn would lead to more generation of plastic wastes. To avoid this vicious cycle, it is necessary to discover safer substitutes to existing plastic additives, utilize biomass and biomass-based plastics, and practice proper management of plastic wastes.

References

1. Geyer R, Jambeck JR, Law KL. Production, use, and fate of all plastics ever made. Sci Adv. 2017;3(7):e1700782.
2. Jambeck JR, Geyer R, Wilcox C, Siegler TR, Perryman M, Andrady A, et al. Plastic waste inputs from land into the ocean. Science. 2015;347(6223):768–71.

3. Sugiura M, Takada H, Takada N, Mizukawa K, Tsuyuki S, Furumai H. Microplastics in urban wastewater and estuarine water: Importance of street runoff. Environ Monit Contaminants Res. 2021;1:54–65.
4. Isobe A, Kubo K, Tamura Y, Kako SÄ, Nakashima E, Fujii N. Selective transport of microplastics and mesoplastics by drifting in coastal waters. Mar Pollut Bull. 2014;89(1–2):324–30.
5. Moore CJ, Moore SL, Leecaster MK, Weisberg SB. A comparison of plastic and plankton in the North Pacific central gyre. Mar Pollut Bull. 2001;42(12):1297–300.
6. Eriksen M, Lebreton LCM, Carson HS, Thiel M, Moore CJ, Borerro JC, et al. Plastic pollution in the world's oceans: more than 5 trillion plastic pieces weighing over 250,000 tons afloat at sea. PLoS One. 2014;9(12):e111913.
7. Matsuguma Y, Takada H, Kumata H, Kanke H, Sakurai S, Suzuki T, et al. Microplastics in sediment cores from Asia and Africa as indicators of temporal trend in microplastic pollution. Arch Environ Contam Toxicol. 2017;73(2):230–9.
8. Koro M, Takada H, Boonyatumanond R, Kwan C. Analysis of microplastic pollution history in Asian countries by using sediment cores. Annual Meeting of Japan Society of Water Environment, Kyoto; 2021.
9. Crutzen PJ, Stoermer EF. The Anthropocene. Global Change Newsletter. 2000;41:17–8.
10. Rothstein SI. Plastic particle pollution of the surface of the Atlantic Ocean: evidence from a seabird. Condor. 1973;75(344):5.
11. Ryan PG. Ingestion of plastics by marine organisms. In: Takada H, Karapanagioti HK, editors. Hazardous chemicals associated with plastics in the marine environment. Cham: Springer International Publishing; 2019. p. 235–66.
12. Tanaka K, Yamashita R, Takada H. Transfer of hazardous chemicals from ingested plastics to higher-trophic-level organisms. In: Takada H, Karapanagioti HK, editors. Hazardous chemicals associated with plastics in the marine environment. Cham: Springer International Publishing; 2019. p. 267–80.
13. Azevedo-Santos VM, Gonçalves GRL, Manoel PS, Andrade MC, Lima FP, Pelicice FM. Plastic ingestion by fish: a global assessment. Environ Pollut. 2019;255:112994.
14. Tanaka K, Takada H. Microplastic fragments and microbeads in digestive tracts of planktivorous fish from urban coastal waters. Sci Rep. 2016;6:34351.
15. Wright SL, Thompson RC, Galloway TS. The physical impacts of microplastics on marine organisms: a review. Environ Pollut. 2013;178:483–92.
16. Kögel T, Bjorøy Ø, Toto B, Bienfait AM, Sanden M. Micro- and nanoplastic toxicity on aquatic life: determining factors. Sci Total Environ. 2020;709:136050.
17. Isobe A. Percentage of microbeads in pelagic microplastics within Japanese coastal waters. Mar Pollut Bull. 2016;110(1):432–7.
18. Everaert G, Van Cauwenberghe L, De Rijcke M, Koelmans AA, Mees J, Vandegehuchte M, et al. Risk assessment of microplastics in the ocean: modelling approach and first conclusions. Environ Pollut. 2018;242:1930–8.
19. Andrady AL, Rajapakse N. Additives and chemicals in plastics. In: Takada H, Karapanagioti HK, editors. Hazardous chemicals associated with plastics in environment. The handbook of environmental chemistry. Berlin, Heidelberg: Springer; 2017. p. 1–17.
20. Tanaka K, Takada H, Ikenaka Y, Nakayama SMM, Ishizuka M. Occurrence and concentrations of chemical additives in plastic fragments on a beach on the island of Kauai. Hawaii Mar Pollut Bull. 2020;150:110732.
21. Yeo BG, Takada H, Yamashita R, Okazaki Y, Uchida K, Tokai T, et al. PCBs and PBDEs in microplastic particles and zooplankton in open water in the Pacific Ocean and around the coast of Japan. Mar Pollut Bull. 2020;151:110806.
22. Tanaka K, Takada H, Yamashita R, Mizukawa K, Fukuwaka M-A, Watanuki Y. Facilitated leaching of additive-derived PBDEs from plastic by seabirds' stomach oil and accumulation in tissues. Environ Sci Technol. 2015;49(19):11799–807.
23. Sun B, Hu Y, Cheng H, Tao S. Releases of brominated flame retardants (BFRs) from microplastics in aqueous medium: kinetics and molecular-size dependence of diffusion. Water Res. 2019;151:215–25.

24. Tanaka K, Takada H, Yamashita R, Mizukawa K, Fukuwaka M-A, Watanuki Y. Accumulation of plastic-derived chemicals in tissues of seabirds ingesting marine plastics. Mar Pollut Bull. 2013;69(1–2):219–22.
25. Tanaka N, Mizukawa K, Takada H, Fujita Y. Uptake and metabolism of plastic-derived chemicals by organisms on sandy beach of Okinawa. Annual meeting of Japan Society of Water Environment, Kyoto; 2021.
26. Tanaka K, Watanuki Y, Takada H, Ishizuka M, Yamashita R, Kazama M, et al. In vivo accumulation of plastic-derived chemicals into seabird tissues. Current Biol. 2020;30(4):723–8.e3.
27. Lavers JL, Hutton I, Bond AL. Clinical pathology of plastic ingestion in marine birds and relationships with blood chemistry. Environ Sci Technol. 2019;53(15):9224–31.
28. Carson R. Silent Spring. Cambridge, MA: Riverside Press; 1962.
29. Cobellis L, Colacurci N, Trabucco E, Carpentiero C, Grumetto L. Measurement of bisphenol a and bisphenol B levels in human blood sera from healthy and endometriotic women. Biomed Chromatogr. 2009;23(11):1186–90.
30. Mori C. Multiple pollution of fetus: Chuko shinsho; 2002.
31. Levine H, Mindlis I, Swan SH, Martino-Andrade A, Jørgensen N, Mendiola J, et al. Temporal trends in sperm count: a systematic review and meta-regression analysis. Hum Reprod Update. 2017;23(6):646–59.
32. vom Saal F, Cohen A. How toxic chemicals contribute to COVID-19 deaths. Environmental Health News; 2020.
33. Wu Q, Coumoul X, Grandjean P, Barouki R, Audouze K. Endocrine disrupting chemicals and COVID-19 relationships: a computational systems biology approach. Environ Int. 2020; In Press
34. Teuten EL, Saquing JM, Knappe DRU, Barlaz MA, Jonsson S, Bjorn A, et al. Transport and release of chemicals from plastics to the environment and to wildlife. Philos Trans R Soc B-Biological Sci. 2009;364(1526):2027–45.
35. Kwan C, Takada H, Mizukawa K, Torii M, Koike T, Yamashita R, et al. PBDEs in leachates from municipal solid waste dumping sites in tropical Asian countries: phase distribution and debromination. Environ Sci Pollut Res. 2013;20(6):4188–204.
36. Kwan CS, Takada H. Release of additives and monomers from plastic wastes. In: Takada H, Karapanagioti HK, editors. Hazardous chemicals associated with plastics in the marine environment. Cham: Springer International Publishing; 2019. p. 51–70.
37. Gomi M, Mizukawa K, Takada H. Development of analytical method based on direct derivatization and estimation of BPA source to Tamagawa river water. 28th Annual meeting of Japan Society of Environmental Chemistry, Saitama; 2019.
38. Bell L, Takada H. Plastic waste management hazards. International Pollutants Elimination Network (IPEN), 2021. ISBN: 978-1-955400-10-7. https://ipen.org/sites/default/files/documents/ipen-plastic-waste-management-hazards-en.pdf.

Chapter 11
Solution and Agreement in the Nuclear Disaster: Toward an Inclusive Society Respecting Relationship from Sacrificial System

Akihiko Kondoh

Abstract Ten years have passed since the accident of Tokyo Electric Power Company Fukushima Daiichi Nuclear Power Plant following the 2011 off the Pacific coast of the Tohoku Earthquake. The release of radioactive materials into the environment resulted in the consequent long-term evacuation of residents. As of 2021, return has been realized in some former evacuation order areas. This paper describes the progress during this period using the Yamakiya district in Kawamata town, Fukushima Prefecture, Japan, which had been designated as an evacuation area as an example. Over the past decade, a variety of stakeholders, including scientists, have been involved in this issue, but differences in positions have also led to differences in the way issues are perceived. It can be said that this deepened the discussion on the relationship between science and society and the relationship between urban and rural areas. However, it is far from solving problems at present. Sustainable Development Goals (SDGs) were established by the United Nations in 2015, and operations aimed at realizing it have begun worldwide. Future Earth Program is also underway to support SDGs from a scientific perspective. In order to achieve SDGs and to transform societies, we need to mourn the victims of the nuclear power plant accident, which is the calamity of modern civilizations, and use the suffering caused by the accident as a lesson.

Keywords Fukushima · Yamakiya district · Nuclear disaster · Inclusive society Science and society · Stakeholders

A. Kondoh (✉)
Center for Environmental Remote Sensing, Chiba University, Chiba, Japan
e-mail: kondoh@faculty.chiba-u.jp

T. Nakajima et al. (eds.), *Overcoming Environmental Risks to Achieve Sustainable Development Goals*, Current Topics in Environmental Health and Preventive Medicine, https://doi.org/10.1007/978-981-16-6249-2_11

11.1 Introduction

The 2011 off the Pacific coast of Tohoku Earthquake, which occurred on March 11, 2011, caused a huge tsunami on the east coast of the Japanese Archipelago and caused the loss of electric power at the Fukushima Daiichi Nuclear Power Plant (hereinafter referred to as "FDNPP") of Tokyo Electric Company (TEPCO) located near the coast. As a result, a steam explosion occurred in the nuclear power plant building, and an unprecedented situation occurred in which a large amount of radioactive material was released into the environment. Major reports on the accident have been submitted by the government [1], the Diet [2], TEPCO [3], and the private sector [4]. The nuclear disaster deprived people of their places of living and led to the emergence of *anoecumene, non-resident areas of humans,* in a modern civilized nation.

Ten years or more have passed since the accident occurred. Except in some areas, evacuation orders have been lifted and people have begun to return. However, it does not mean a solution to the problem. It was a bitter decision made by local governments who wanted to avoid disputes with the national government and residents who wanted to return to their hometowns as soon as possible, and it was an inevitable decision. It is not the result of the restoration of trust between TEPCO, which caused the accident, and the government, which promoted the nuclear policy, and the people affected by the accident.

Nuclear power generation is a state-of-the-art technology that gathers excellence from science and technology, and has made a major contribution to the establishment of a modern civilized state. However, it must not be forgotten that there are many people who have been sacrificed behind the scenes.

In retrospect, behind Japan's high economic growth was the calamity called pollution. Though there were *Minamata disease*, *Itai-itai disease*, *Niigata Minamata disease,* and *Yokkaichi air pollution* as four major pollution diseases in Japan, there was the efficient way of the industrial production which did not consider the environment in the back. Behind the benefits of economic growth for many citizens, groups of victims were forced to fight long-term patience.

This means that Japan's development has been achieved on the "sacrificial system" [5]. Sustainable Development Goals (SDGs) [6] are currently underway worldwide. The goal of SDGs is social transformation. Actions corresponding to 17 goals with the goal of *not leaving anyone behind* are underway with the aim of achieving the goal in 2030. The goals of SDGs are based on various stakeholder coordination and compromise as administrative documents, but behind them are the actual circumstances of the United Nations (UN) participants facing real issues. Behind a sentence, there are truly thoughts that the current problems are solved and bright future will be available. What is social change to achieve SDGs? World citizens need to continue to think sincerely.

Under these circumstances, a pandemic caused by COVID-19 occurred. Its end is still at an unforeseen stage, but the world will eventually overcome the corona outbreak. So is post-Corona society a return to the previous world? SDGs must make the post-corona community a new, reformed society today with the goal of social change.

11.2 Course of Nuclear Disaster

11.2.1 Summary of the Overall Course

The Tsunami accompanied by The 2011 off the Pacific coast of Tohoku Earthquake, caused flooding damage to the FDNPP, and the reactor became uncontrollable due to the loss of electric power. On March 12, the reactor building exploded one after another, and radioactive materials were released into the environment. The radioactive plume released on the afternoon of March 15 flowed northwestward, and a large amount of radioactive material was deposited on the Abukuma highlands together with the spring snow. This deposition will set up evacuation areas not only near the FDNPP but also in areas more than 40 km away from the FDNPP [1–4].

Immediately after the nuclear accident, in order to know the radioactivity (spatial dose rate) distributions over a wide area, Japanese government conducted a spatial dose rate survey as a joint project of Department of Energy (DOE) in the United States and Ministry of Education, Culture, Sports, Science and Technology (MEXT). It is the so-called aircraft monitoring. The first measurement was made on March 17, and the results were posted on the DOE website on March 22 and communicated to the field through researchers and supporters. Aircraft monitoring in the 80 km radius of the FDNPP by Japan was carried out several times, and a spatial dose rate map was developed. Based on this map, a deliberate evacuation area was established (the first map was published by the Japanese government on May 8, 2011). Subsequently, the national government continued to measure the spatial dose rate by running monitoring, monitoring posts, etc.

Radioactivity monitoring by researchers was performed in parallel with national radioactivity monitoring. The objectives were to quickly convey information to the region and to clarify the distribution of radioactivity in the spatial region involved in human life. In the Yamakiya district, the air dose rate distribution in the forest area was measured and periodically reported to the area [7]. There were many other researchers who entered the community and acted with the residents in an effort to restore the environment and livelihood.

In this way, researchers had divided into two groups, the one considered the role of clarifying the actual condition of radioactive contamination and the transition of radioactive materials in the environment, and the other was a group closely related to the region and aimed at solving problems in the region. The difference between these two groups was the perception of stakeholders. The stakeholder of former group is a nation or a world, and it was regarded as an obligation to publish elucidated actual condition to the world in the form of scientific paper. The latter regarded the problem as a social problem to be solved, and assigned the party of the field as stakeholder. Stakeholders are diverse and hierarchical, and there can be conflicts among stakeholders. Though the action which aims at the solution of the problem which is actually occurring, also has the hierarchical structure itself, the decoupling among the hierarchies was also recognized.

Many papers covering FUKUSHIMA from a scientific perspective have been published in journals worldwide. On the other hand, although there are many papers written in Japanese on the progress of efforts in the field of FUKUSHIMA, please refer to Yamakawa and Yamamoto eds. [8, 9] as a book disseminated worldwide in English.

11.2.2 Progress in the Yamakiya District

The author has carried out various activities in the field of Yamakiya district, Kawamata town, Date District, Fukushima Prefecture, which was an evacuation area. I would like to explain the Yamakiya district as an illustration for knowing FUKUSHIMA site. Although the Yamakiya district is a part of Kawamata town, it was decided that only the Yamakiya district would be designated as an evacuation area based on the distribution of the spatial dose rate. Located on the highlands of Abukuma Highland, the center of the town was located in a basin in the northwest direction of the Yamakiya district, and the administrative functions as a town were maintained.

After the nuclear power plant accident became apparent, the Chief Cabinet Secretary's words "*There is no immediate harm to health*" were repeatedly broadcast on television. It was an advice based on scientific rationality, not a wrong message. Around that time, in the Yamakiya district, farm work and preparations for the new semester had begun amid anxiety. On April 12, the national government announced that a systematic evacuation would be carried out, and an evacuation order was issued on April 22. However, the evacuation was not systematically carried out, and the response of residents began with the search for evacuation destinations. In fact, they spent about three months preparing for evacuation and were forced to live in radiation during that time [10]. The situation was similar in neighboring municipalities.

Two years after the evacuation has been almost completed, the Yamakiya district has reached an agreement to review the evacuation area, and on August 8, 2013, it will be reorganized into the evacuation order lifting reserve area and the residential restriction area. This was the last reorganization among the evacuation municipalities in the vicinity, because there was a conflict in the district due to the difference in the degree of spatial radioactive contamination. Against this backdrop, in December 2014, people wishing to return home established the *Yamakiya Easy Network*, non-profit organization (NPO), and began its activities. On the other hand, alternative dispute resolution (ADR) lawsuits have been filed, and the idea of homage has been complex and diverse.

The "*Verification Committee on Decontamination of Yamakiya District*" was launched in April 2015, when other local governments had seen the movement to lift evacuation orders, and the author also decided to participate as a member of the Committee. Since the decontamination project undertaken by the national government is to be completed by December 2015, the committee's objectives were to summarize: the analysis and verification of the decontamination effect of the Yamakiya district; the investigation and research results on radiation in the Yamakiya district; and the restoration of the environment in the Yamakiya district. In the interim report summarized in July, 2015, based on the completion of decontamination business around housing land and crop land, this paper analyzes, verifies, and evaluates the effect, etc., and summarizes the proposal on matters related to decontamination and radiation protection measures, etc.

Items concerning the decontamination of residential land and the evaluation of health effects in the interim report were conducted based on the annual additional exposure dose of 20 mSv, which is the national decontamination target. As a result of the evaluation based on official data authorized by the government, i.e., the results of measurement of the spatial dose rate around the decontaminated housing land, the latest records of monitoring posts and running surveys, etc., the annual exposure dose was less than 20 mSv and the achievement of the target value was confirmed. Of course, the report described the need for ongoing monitoring because there are many sites where the long-term target for the annual additional exposure dose, 1 mSv, is exceeded.

There are various different concepts for the annual additional exposure dose of 20 mSv. The committee was to make "scientific" decisions, but currently only 100 mSv have the scientific basis for the annual additional exposure dose for carcinogenesis. 20 mSv is the recommendation of the International Commission on Radiological Protection (ICRP), and the concept behind it is the Threshold-Free Linear (NLT) hypothesis. Since the national decontamination target is 20 mSv based on ICRP recommendations, it was decided that the role of the decontamination verification committee was to confirm annual additional dose rate was lower than 20 mSv. After confirmation, the next step begins, but the 20 mSv is no longer a reference for return, and it is not possible to "scientifically" determine its value. Nonetheless, a certain level must be determined, so that it is an "appreciation level." Since doses exceeding the long-term target of 1 mSv are accidental exposures, TEPCO and national government may be able to determine the level of understanding by presenting their benefits in balance with the risks associated with additional exposure doses. However, local governments hesitate to confront national government with policies and budgets, and residents who hope to return hesitate to take time to reach an agreement. Originally, the person responsible for the nuclear disaster should apologized at first, and while the government, TEPCO, and the victims would jointly discuss the future of the region, the level of understanding would be determined. However, the hierarchy of stakeholders, conflicts among stakeholders, and other factors hindered the establishment of the level of understanding. There may be many levels of understanding, as there are different circumstances for each evacuation municipality. While respecting various decisions and promoting double-track reconstruction, the desire to return should have sought a level of understanding, but the unified desire of national government has been prioritized.

Recommendations in the latter half of the report describe the preparation of maps for realistic radiation protection in existing exposure situations, measures against radiation in forests where people enter on a daily basis, a comprehensive counselor system, accelerated environmental recovery in daily life, and measures to deal with domestic and overseas markets. These recommendations do not necessarily lead to the immediate realization of concrete measures, but they are regarded as equivalent to the "Basic Law" in the law, and they describe the philosophy and basic policy for bridging future measures corresponding to the "Individual Law." Collaboration with experts has already begun with a view to realizing this goal, and applications for subsidies for support activities and discussions are being held.

11.2.3 Scientists' Behavior

When the FDNPP accident revealed the release of radioactive material into the environment, scientists began to act. At this time, there seem to have been two behaviors of scientists as mentioned above. One is an attempt to clarify the mechanism of the transfer of radioactive material from a scientific standpoint. The other was to enter the field of a nuclear disaster, clarify the actual situation of the damage with stakeholders, and think about the ideal way of the future region.

In the field of hydrology, which is a specialty of the author, there was a growing momentum that the knowledge and experience of hydrology could be utilized for grasping the present state and environmental restoration in the stage in which the actual condition of the damage was clarified. Especially, the problem of what kind of process the radioactive material released and deposited in the environment by FDNPP accident goes into the human dwelling area is regarded as an important problem which the hydrology which handles water and material circulation in the watershed should shoulder. One field of hydrology is the field of slope hydrology. This deals with the challenge of how precipitation on mountain slopes flows through processes into mountain streams and becomes river water. The process of water infiltration and efflux on a vegetated slope is composed of multiple elementary processes, with substantial changes in river water volume and substantial material movements being different. This was an essential knowledge to predict the migration of radioactive material.

The results of surveys and studies on the transfer of radioactive material in watersheds and mountainous slopes were summarized in Onda et al. [11]. It can be said that scientific knowledge has been obtained on the kinetics of the migration of radioactive substances in the environment, in this case cesium-137. The next issue is how to utilize the knowledge obtained in the measures in the field. The results of science alone do not mean that humans can act reasonably. Behavior is decided by the understanding which made various circumstances and thoughts to be a background.

Although the disaster-stricken areas of the nuclear disaster in Fukushima were covered by multiple local governments, some scientists entered the local governments and regions where they were able to build relationships, and attempted to clarify the actual situation of the nuclear disaster from a wide range of viewpoints. The academic fields which became the background were various such as science, engineering, sociology, planning, etc., and they decided to practice interdisciplinary and transdisciplinary.

The team of Chiba University entered the Yamakiya district of Kawamata town, one of the former deliberate evacuation districts, and researchers in various specialized fields such as hydrology, ecology, agricultural economics, and nursing have continued activities related to comprehensive environmental monitoring and regional reconstruction in cooperation with local residents and municipalities. Part of the progress is summarized in Kondoh and Hama [12].

11.2.4 Consent to the Exposure

Immediately after the accident, MEXT published supplementary reading book for students in elementary, middle, and high school on radiation to disseminate knowledge about radioactivity. In the inside, the radiation is divided into artificial radiation and natural radiation, and it explains that the artificial radiation has merits and that there is the exposure of natural radiation even in the usual life. So it was the basis for asserting safe feeling under additional dose of 20 mSv/year as a target value of decontamination, but this did not give us any peace of mind.

In the Yamakiya district, test decontamination was carried out in late 2011, and full-scale decontamination work by the national government started in 2012. As a result, the air dose rate in the residential area gradually decreased, and the evacuation order was lifted when it was confirmed that the dose rate fell below the standard. On the basis of scientific rationality, it is concluded that health effects do not appear immediately. However, from the perspective of the disaster victims, there is a third type of exposure. It is a forced exposure in an accident that is not willing to do so. Understanding is straightforward if the traditional science divides the sense of value and mind with the object, makes a third-party judgment, and gives the victim a spiritual attitude to accept it. However, it is not possible to underestimate the forced exposure due to the accident, because the object is human and the object with the heart.

Society's use of modern science-based technologies such as nuclear power generation requires a mechanism to understand the benefits and risks of these technologies and not to divide them. This is a lesson learned in Fukushima and a challenge related to the achievement of the 12th goal of SDGs: *Responsible Production and Consumption*. If there is no relationship of trust between the position of *Production* electricity using nuclear energy and the position of *Consumption* electricity, the technology should be considered not to be in the stage of use.

11.2.5 Restoration of Mountain Villages

Ten years have passed since the accident in March 2011. The extent of the nuclear disaster varies from region to region, and the state of recovery and reconstruction is also diverse. Although 364 households and 1252 persons were forced to evacuate in the Yamakiya district of Kawamata town, where the author was involved, only 343 people reside in the Yamakiya district as of January 2021. Although the number of residents was less than 30% of the number before the accident, the number of registered residents was 727, and half of the residents were unable to decide on the end [13].

The *Report of the Verification Committee on Decontamination of Yamakiya District, Kawamata Town* [14], which was prepared as a procedure for lifting evacuation orders, made the following five recommendations for regional revitalization in Yamakiya District.

1. Map for realistic radiation protection.
2. Radioactivity measures for forests in which people are entering on a daily basis.
3. Consultation response17.
4. Accelerating the recovery of the environment for everyday life.
5. Measures for domestic and overseas markets.

Among them, some of the measures for (3) and (4) have been taken, but less progress has been made for other items. The return of some residents has already begun, but the district has become an aged-centered society, and the meaning of living in radioactive contaminated areas has changed drastically now that many young people have had the basis for their lives outside the district. It can be said that for returnees, the joy of living in their hometowns surpassed the suffering. Nevertheless, the future of the area affected by the nuclear disaster, including the Yamakiya district, is still chaotic.

11.2.5.1 Maps for Radiation Protection

There are two purposes for mapping the spatial dose rate. The spatial dose rate mapping performed by MEXT/DOE immediately after the accident was the monitoring required for drawing the evacuation area. However, it seems that the image that the wide area is uniformly contaminated by the map of the reduction scale was given except for Fukushima residents. However, the distribution of the spatial dose rate of the large-scale map was highly heterogeneous depending on the migration path of the radioactive plume, the topography of each location, and the geographical conditions such as vegetation, as described in Kondoh and Hama [12]. When feedback enters the field of view, it is necessary to know the spatial dose rate or the distribution of radioactivity on the scale of human life. This requires monitoring in accordance with the way of living in polluted mountain villages. In Kondoh and Hama [12], detailed spatial dose rate distributions were obtained by walking in areas including mountains and shared with local residents. It gave hope in some places, but it was sometimes used as a basis for giving up.

The map of the spatial dose rate from the walking survey was intended to be used by residents who wanted to return to their homes for future planning, but it was not really a top priority for residents who were evacuating. The most important issue was the revitalization of the livelihood. We cannot return to our hometowns if our lives do not exist. On the other hand, after the evacuation order was lifted and the return began, the distribution of the spatial dose rate was no longer an important concern. Of course, mountain forests exist behind the mountain village settlements, and the current situation of the spatial dose rate in the vicinity needs to be known, but the areas where monitoring is necessary have been limited. Kondoh [15] proposed the GIS improvement aiming at the decontamination of the *SATOYAMA* basin unit. *SATOYAMA* is an area where people and nature interact. Life in mountain villages is essential to the use of ecosystem services in *SATOYAMA*, and is a source of pride for living in mountain villages. The purpose of GIS construction was to describe the geographical attributes of the Satoyama basin, including the settlements and the *SATOYAMA* behind them, in GIS, and to contribute to radioactivity management, but it was also intended as a visualization tool that links the divided urban world with the world of mountain villages.

Thus, scientific monitoring of radioactive materials was necessary, but the priority was never high among the livelihoods in mountain villages that had not yet recovered. This is also an important issue concerning the relationship between the problems to be solved and science and technology. This means that *science in a narrow sense is necessary to solve problems, but science alone cannot solve problems*, which is called trans-science.

11.2.5.2 Forest Radioactivity Countermeasures

During the evacuation order, government-managed decontamination was carried out, and the committee authorized that the spatial dose rate in residential areas and fields declined and that the annual additional exposure dose was less than 20 mSv, which is the standard for lifting evacuation orders. However, radioactive materials in forests remain there. Is it necessary to take measures against radioactivity in forests?

Living in cities is separated from nature, and there are few opportunities to directly gain blessings from nature. In Japanese mountain villages, however, the wild vegetables and mushrooms obtained from forests are important resources that color the livelihoods, and they are positioned as minor subsistence in their daily lives. However, minor subsistence was not included in the compensation framework for the accident [16]. It is because we cannot recognize the necessity of life in the urban world. Approximately 90% of the population in Japan is urban. On the other hand, about 70% of the national land is occupied by forests, which is a place for living in mountain villages. Diverse values should be respected and measures against radioactivity in forests should be considered. On September 9, 2012, the Japan Afforestation

Engineering Association adopted the *Appeal on Pollution Countermeasures in SATOYAMA Areas Damaged by Nuclear Disasters* and sent it to relevant administrative agencies. At present, it is difficult to solve realistic problems, so radioactivity countermeasures in mountain forests have not been carried out, but it seems to have been able to cultivate understanding about the life of mountain villages.

11.2.5.3 To Accelerate the Recovery of the Environment through Consultation and Daily Life

Various administrative support measures have been implemented in this area. The menu is also available on Kawamata Town's website, and a number of measures are presented regarding support for return, resumption of business, and countermeasures against radioactivity.

As a reconstruction base in the Yamakiya district, the restoration base commercial facility, *Tonya no Sato,* opened in July 2017. This facility assumes the life support and exchange function of returnees, and it aims at contributing to the regeneration of the local community. In addition, public facilities are located in the vicinity, and they aim to function as a base in the region. In this way, the aspect in which the base inspires the activity in the region is important, and it becomes the basis of the activity of the inhabitant for the reconstruction of the beautiful mountain village. Nevertheless, there are many difficulties, such as NPOs taking over operations after the withdrawal of restaurant operators, but living in hometowns gives courage to people.

11.2.5.4 Measures for Domestic and Overseas Markets

Recovery of the market lost by the accident is the top priority for achieving recovery and reconstruction of life. However, rumor damage makes it difficult to achieve the problem. Although agricultural and marine products produced in the former evacuation zone have been thoroughly inspected, the market has not completely recovered at present. This means that consumers do not decide to buy only with scientific rationality. The problem is not solved when the scientific communication to the consumer is achieved, and it means that the understanding of the human side of the disaster recognition needs to be promoted. Although it is an issue for the future, it is also an important issue related to the way modern civilization should be.

11.2.5.5 Yamakiya's Future

Since activities to lift the evacuation order must give priority to restoring the pride of people's livelihoods and local communities, efforts were made to consider the future of the region within the framework of greater people-to-people relations. Around this time, the number of citizens and scientists outside the region related to the Yamakiya district also increased, and the phase of the activity has also changed from the activity of the investigation main body in the early stage to the activity for the restoration of living industry and the construction of the social relation capital. Yasutaka et al. [17] point out the importance of dialogue and volunteer activities in the reconstruction of local communities in addition to surveys in activities after the lifting of evacuation orders. Nevertheless, the way of thinking about the region's future varies. The possible path to reconstruction is the process in which stakeholders, including researchers, participate in activities to create the future of the community and form consensus with confidence.

11.3 Final Remarks

After the period of high economic growth in late twentieth century, Japan entered a stage of low growth or steady society. The situation is similar in many developed countries around the world. Now that we have experienced FUKUSHIMA and COVID-19, we must review how modern civilizations should be with a sincere attitude, because it is thought to be predisposed to scientific and technological developments in the twentieth century, such as nuclear power in the case of Fukushima and aircraft, facilitating the movement of people, in the case of COVID-19. For that purpose, we must understand the present ideal way based on the history from the past, and look forward to the future while trying to solve the present problems. The present must not be ignored when talking about the future. This is an achievement of SDGs.

People want to talk about ideals when it comes to the future. However, the future is connected to the present, and we cannot tell the future with the solution of the present problem on hold. Do not neglect remembering the devastation of FUKUSHIMA and continuing to question its implications.

References

1. Investigation committee on the accident at the Fukushima nuclear power stations. http://www.cas.go.jp/jp/seisaku/icanps/eng/.
2. The national diet of Japan Fukushima nuclear accident independent investigation commission. https://warp.da.ndl.go.jp/info:ndljp/pid/3856371/naiic.go.jp/en/.
3. TEPCO nuclear accident independent investigation commission. https://www.tepco.co.jp/cc/press/2012/1205628_1834.html.
4. The 10-year investigation commission on the Fukushima nuclear accident publishes final report on lessons learned since the disaster. https://apinitiative.org/en/2021/02/10/16218/.
5. Kondoh A. Solution and agreement in the nuclear disaster: toward an inclusive society respecting relationship from sacrificial system. Trend Sci. 2019;24(10):49–52. (in Japanese)
6. United Nations. Transforming our world: the 2030 Agenda for Sustainable Development. 2015. https://www.mmv.org/sites/default/files/uploads/docs/publications/2030_Agenda_Sustainable_Development.pdf.

7. Kondoh A, Kobayashi T, Kinoshita I, Yamaguchi H, Hayakawa T, Matsushita R. Report on the spatial dose rate and inventory in Kawamata Town. J Rural Plan Assoc. 2011;30(3):419–20. (in Japanese)
8. Yamakawa M, Yamamoto D, editors. Unravelling the Fukushima disaster. London: Routledge; 2016.
9. Yamakawa M, Yamamoto D, editors. Rebuilding Fukushima. London: Routledge; 2017.
10. Kanno G. Thinking from planned evacuation area: Go forward to return. Mod Agric. 2012;2012(7):360–3. (in Japanese)
11. Onda Y, Taniguchi K, Yoshimura K, et al. Radionuclides from the Fukushima Daiichi Nuclear Power Plant in terrestrial systems. Nat. Rev. Earth Environ. 2020; https://doi.org/10.1038/s43017-020-0099-x.
12. Kondoh A, Hama A. Chapter 17 Nuclear disaster and human geoscience. In: Himiyama Y, Satake K, Oki T, editors. Human Geoscience. Springer; 2019. p. 339.
13. Kawamata Town. Residence status in Yamakiya District. (in Japanese) https://www.town.kawamata.lg.jp/site/sinsai-saigai/yamakiyatikukyojyuujyoukyou.html
14. Kawamata Town. Verification committee on decontamination for Yamakiya District. (in Japanese) https://www.town.kawamata.lg.jp/site/sinsai-saigai/kawamatamatiyamakiyatikujo-senntounikakannsurukennsyouiinnkai.html
15. Kondoh A. Construction of geographic information system aimed at radioactivity measures in the *SATOYAMA* watershed. J Japan Soc Revegetation Technol. 2012;28(2):274–7. (in Japanese)
16. Kaneko H. Radioactive contamination of mountains and its damage to local communities: from the view point of miner subsistence. J Environ Sociol. 2015;21:106–21. (in Japanese with English Abstract)
17. Yasutaka T, Kanai Y, Kurihara M, Kobayashi T, Kondoh A. Takahashi T, and Kuroda Y dialogue, radiation measurements and other collaborative practices by experts and residents in the former evacuation areas of Fukushima: a case study in Yamakiya District, Kawabata Town. Radioprotection. 2020;55(3):215–24.

Chapter 12
Current Environmental Risks and their Management in Japan

Teruhisa Tsuzuki and Tamie Nakajima

Abstract The Joint Open Symposium of the Science Council of Japan and the Japan Society for Occupational Health, entitled "Environmental Cycling of Toxic Materials and Health to Archive Sustainable Development Goal (SDGs) 12 Responsible the consumption, and production-" was held at the 92nd Annual Meeting of the Japan Society for Occupational Health on May 25th, 2019. At this symposium, four researchers lectured on environmental risks we are currently facing, as described in Chaps. 8–11. It was emphasized that regulatory science has implications on every aspect of human life.

Keywords Asbestos · Nuclear disaster · Plastic problem · Polychlorinated biphenyls · Risk assessment · Risk communication · Risk management

12.1 Proper Management of Toxic Chemical Substances and Problems

Many chemical substances overflow our lives today. We enjoy the benefits and convenience of so many chemical substances in our daily lives. However, these chemical substances, including intermediates, produced during the manufacturing process

T. Tsuzuki
Department of Medical Biophysics and Radiation Biology, Faculty of Medical Sciences, Kyushu University, Fukuoka, Japan
e-mail: tsuzuki.teruhisa.202@m.kyushu-u.ac.jp

T. Nakajima (✉)
Nagoya University, Nagoya, Japan

College of Life and Health Sciences, Chubu University, Kasugai, Aichi, Japan
e-mail: tnasu23@med.nagoya-u.ac.jp

T. Nakajima et al. (eds.), *Overcoming Environmental Risks to Achieve Sustainable Development Goals*, Current Topics in Environmental Health and Preventive Medicine, https://doi.org/10.1007/978-981-16-6249-2_12

may have adverse effects on the human body as well as the ecosystem. Stockholm Convention on POPs was adopted in 2001, which came into effect in 2004. In Japan, polychlorinated biphenyl (PCB), which is a representative of POPs, was initially used because of its stability and excellent electrical insulation properties. PCB was accidentally mixed in the cooking oil production process in 1968 and caused health problems to many people (Yusho patients). As a result of this poisoning case, its manufacture and use were banned in 1972. In 2001, the "PCB Special Measures Law" was enacted, and the processing business is in progress. However, the construction of the treatment facility was not welcomed by the residents. Moreover, as Dr. Masunaga pointed out (Chap. 8), individual regulations for drugs, pesticides, and cosmetics must be performed. To improve the current situation that lacks a comprehensive legal system for handling chemical substances and an environmental risk assessment system, we should urgently establish a "risk–benefit comparison" system for alternative substances. The "Law Concerning Chemical Substance Examination and Manufacturing Regulations" was enacted to prevent environmental pollution due to chemical substances in 1973, which was revised in 2009, and the examination and management are in progress.

12.2 Health Risk Assessment and Management of Asbestos Exposure to Citizens

Asbestos has been widely used in buildings as a heat-insulating material and soundproofing material, but it causes health problems, such as lung cancer and mesothelioma. Although it was banned in 2012, still, many buildings contain asbestos. Therefore, measures to prevent asbestos scattering should be taken when dismantling buildings. However, asbestos leakage accidents due to inadequate leakage prevention measures have been reported. Dr. Hisanaga introduced (Chap. 9) the efforts of the expert review group for risk assessment following a leak accident that occurred during the asbestos removal work at a subway station in Nagoya in 2013. The group has recommended recurrence prevention and health measures based on the spread of asbestos from the station, exposure of station users and workers, and their health risk assessment. Other than this leak, citizens may be exposed to asbestos from serpentinite, so it is necessary to know the actual chemical exposed. Since there are currently few human resources in such fields, it is necessary to educate personnel who can oversee the risk management of chemical substances.

12.3 Solving the Marine Plastic Problem

Plastic production increases at a rate of 5% every year, and the world's production exceeds 400 million tons per year. About half of it becomes disposable, such as plastic bottles, plastic bags, and food packaging. The plastics that flow into the sea form microplastics due to ultraviolet rays and wave forces. Based on the ecological research on seabirds, Dr. Takada has found that the toxic effects of microplastics include their physical effects, as well as additives used in plastic manufacturing and harmful chemical substances, such as DDT and PCB adsorbed on microplastics (Chap. 10). We addressed that the basic requirements for solving the plastic waste problem include 3R—reduce, reuse, and recycle. Among them, he emphasized that "Reduce," which is the control of plastic waste generation, has the highest priority to reduce the amount of plastics that flow into the sea.

In the "leading declaration" of the 20 major countries/regions (G20) summit meeting held in Osaka in June 2019, Japan Prime Minister aimed to reduce the new pollution of marine plastic to zero by 2050. "Osaka Blue Ocean Vision" was included. However, the existing plastics remain to be handled. We should realize the importance of reducing the production and use of disposable plastics toward SDG Goal 12 (Responsible consumption, production).

On April 7, 2020, the Environmental Risk Subcommittee of the Science Council of Japan proposed that industry, government, and private sector should take action to develop technology for promptly recovering and processing existing marine plastics and reduce the amount of disposable plastics.[1]

12.4 Risk Communication Learned from Nuclear Disaster

Due to the large-scale and wide-area radioactive material contamination from the Tokyo Electric Power Company's Fukushima Daiichi Nuclear Power Plant that occurred following the Great East Japan Earthquake on March 11, 2011, many residents were forced to leave their communities for a long time. Dr. Kondoh (Chap. 11) picked up the problem of the evacuation area investigated by the Chiba University after the accident as an example, and gave a lecture on the situation of pollution during and after the evacuation, and introduced transdisciplinary efforts aimed at solving problems. He pointed out that the problem-solving of a nuclear disaster through collaboration with residents is not only "scientific or economic rationality," but requires "shaping of the consensus" based on shared empathy for the region and the idea of how society should be. For Japan to solve the problems deriving from the unprecedented large-scale nuclear disaster and form a sustainable society, as described in Chap. 11, efforts toward an inclusive society that respect diverse values and relationships should be made by multiple stakeholders.

As a whole, the methodology of regulatory science is being established and expanded to include various harmful substances. Among them, the risk communicator still feels immature. Consensus building will not be easy as diverse personal thinking and social backgrounds make it challenging to resolve conflicts.

---

[1] Ecological and Health Effects of Microplastic Pollution of Water Environment Need for Research and Governance of Plastics. Science Council of Japan April 7, 2020 http://www.scj.go.jp (in Japanese) n. Accessed DD Science Council of Japan.

Parts III
Towards the Realization of a Sustainable and Well-Being Society

Chapter 13
The Future of Washing

Youhei Kaneko

Abstract The act of washing keeps our bodies and homes clean and also allows us to prolong the life of our clothing and dishes. Washing is critical for achieving a sustainable society; however, a large amount of resources, including water, energy, and detergents are required. Therefore, it is time to reconsider our washing activities by focusing on more sustainable methods and create the future of washing.

This chapter explores six different approaches for designing and implementing a more sustainable washing life cycle. In order to create the future of washing, a wide range of research and fieldwork needs to be conducted, including: assessing the environmental and social impacts of washing; developing new sustainable washing technologies; and taking into consideration behavioral science. Moreover, key actors, such as government, academia, and the private sector need to come together to discuss the various aspects of washing. The future of washing is essential for our lives and represents an important theme for achieving a sustainable society. It is critical to gather diverse knowledge and ideas from experts and consumers in order to begin discussions about the future of washing.

Keywords Washing · Sustainability · SDGs · Life cycle assessment · Detergent
The future of washing

Y. Kaneko (✉)
Kao Corporation, ESG Division, Tokyo, Japan
e-mail: kaneko.youhei@kao.com

T. Nakajima et al. (eds.), *Overcoming Environmental Risks to Achieve Sustainable Development Goals*, Current Topics in Environmental Health and Preventive Medicine, https://doi.org/10.1007/978-981-16-6249-2_13

13.1 Introduction

People interact with various chemicals in their daily lives when using products that contain synthetic fibers, paints, and plastics. For example, these common chemicals are used in food packaging and home appliances. Detergents are also made from chemicals and are essential not only for cleaning our bodies, but also for clothing, dishes, household items, and furniture.

Washing usually consumes a large amount of water and detergent. Unfortunately, wastewater released in Japan has previously caused environmental issues. For example, in the 1960s, wastewater release created detergent bubbles, in rivers and sewage-treatment facilities, leading to eutrophication in lakes, marshes, and coastal waters. This is thought to be the first incident where people learned how washing could impact the environment. In order to address these issues, the Japanese government improved sewage-treatment facilities, while the business sector pioneered biodegradable detergents and low or non-phosphorus detergents. In addition, environmental damage sparked research into the impacts of detergent on rivers that have carried on to today [1].

13.2 Considering the Life Cycle of Washing

The 1992 Earth Summit held in Rio de Janeiro addressed regional environmental issues, including pollution, climate change, biodiversity, water, deforestation, and chemicals management. Since that time, these issues are now global in nature. Therefore, governments, local communities, and the public are taking action to address these issues. Further, in 2015, the United Nations addressed important social issues, such as human rights, poverty, and conflict, by adopting the Sustainable Development Goals, or SDGs.

Evaluating the environmental impacts of washing from a global, SDG perspective requires us to evaluate the entire life cycle, including procurement of raw materials and disposal of washing goods. In this chapter, the author reviews the life cycle of clothes washing as an example (Fig. 13.1). For example, washing clothes requires the use of detergents, water, and washing machines. The main component of detergents is called a surfactant, which is produced from petroleum or palm oil through a chemical process, requiring both energy and water. Palm oil production in Malaysia and Indonesia requires that rainforests be converted into farmlands, leading to concerns about the impacts on local residents and animals [2]. For example, there are concerns about the working environment and rights of farmers. In addition, water is distributed to households after being treated at water purification plants; however, this process consumes a large amount of energy. Using a washing machine also requires electricity. In addition, some wastewater from clothes washing flows into nature including rivers, impacting animal populations. Additional wastewater also flows into sewage facilities, where again, a great deal of energy is used.

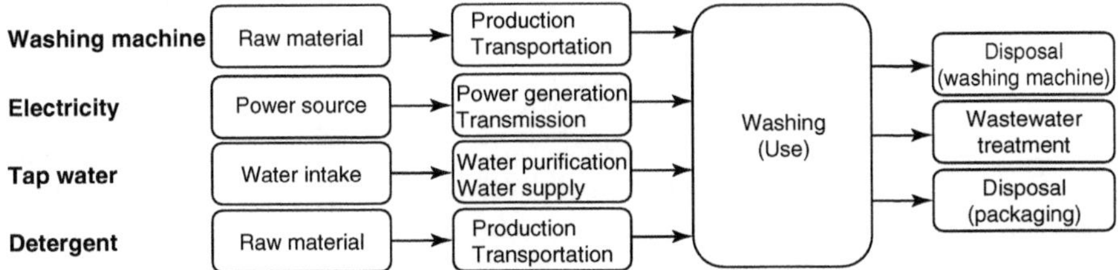

Fig. 13.1 Flows of clothes washing life cycle

The Japan Chemical Industry Association states in its report, *"Innovations in Greenhouse Gas Reductions,"* that the life cycle of CO_2 emissions per kilogram of clothes washing is 0.12 to 0.15 [3]. Data show that people roughly wash 1.5 kg of closes per person a day, meaning around 550 kg of clothing per person is washed annually, which translates into nearly 60 kg of CO_2 emissions. As a result, this means that approximately 6 to seven million tons of CO_2 emissions are generated by Japan's population of 120 million.

Japan's Ministry of Land, Infrastructure and Transport [4] reported that, in 2016, the quantity of domestic water consumption was 12.9 billion m^3. Furthermore, the Tokyo Metropolitan Government Bureau of Waterworks [5] reported that 15% of domestic water use was dedicated to clothes washing. That is, yearly clothes washing in Japan uses approximately 1.9 billion tons of water.

Ohtawa, et al. evaluated the environmental impacts of clothes washing by using LIME2 [6]. The assessment was implemented in 1987 and identified the specific environmental impacts of clothes washing. The report states the following about the environmental effects per clothes washing: 49% contribute to urban air pollution (SO_x, NO_x, etc.); 33% contribute to global warming (CO_2); and there are negative effects on resource consumption and acidification (Fig. 13.2).

As mentioned above, clothes washing is not only related to environmental issues, such as global warming and water scarcity, but also many social ones, such as converting rainforests into palm farming lands.

The world population is projected to reach 10 billion by 2050, increasing the size of the current global economy by three-fold. As a result, clothes washing will likely increase environmental and societal impacts. If measures are not taken, clothes washing may further increase CO_2 emissions and lead to conflicts over water resources. Clothes washing may also lead to an expansion of palm fields, including the associated environmental and societal impacts. There also may be one day when washing is affected by resource scarcity impacting our ability to clean and maintain clothes.

13.3 Considering the Future of Washing

Washing is a critical action to support our daily lives, including cooking, clothing, and housing. It needs to be sustainable and now is the time to think about how the future of washing should look. Below are my insights on the future of washing,

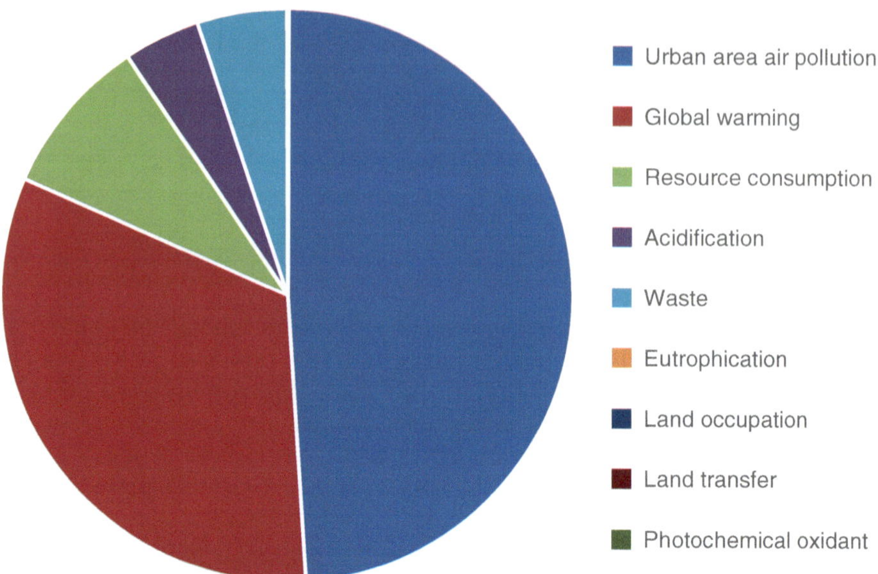

Fig. 13.2 Evaluation of the environmental impacts per clothes washing (1987, detergent: New beads manufactured by Kao Corporation)

which would allow the world to live clean and comfortable without passing on any negative effects to future generations.

13.3.1 Evaluation of the Environmental and Societal Impacts of Washing

The possible environmental and societal impacts of washing have often been discussed. However, these impacts are not measured quantitatively, and thus, the root causes and necessary actions have not been identified and prioritized. It is necessary to make quantitative assessments of the impacts and determine which areas of washing (clothes, bathing, tableware, etc.) and stages of the process (raw materials, water, wastewater, etc.) need to be assessed.

13.3.2 Defining Sustainable Washing

Washing uses critical resources including energy, water, and palm oil. In order to avoid excessive washing and define a concept of sustainable washing, people need to decide how much cleanliness is desired.

13.3.3 Sophistication of Washing Technology

It is critical to develop technologies that can reduce a significant amount of the negative environmental and societal impacts of washing. From 1987 through 2011, Ohtawa et al. assessed the environmental impacts from clothes washing by using LIME2 (Fig. 13.3). More than 65% of the environmental impacts identified in the assessment were actually reduced by 2011. For example, improvements in washing machines and detergents reduced water consumption, power, and detergent quantity. However, further developments in detergents and washing machine technology are needed to reduce the negative impacts of washing.

13.3.4 Changing Consumer Awareness and Behavior

Consumer awareness and behavior need to be transformed in order to promote sustainable clothes washing. For example, possible efforts might be changing clothes washing habits, such as only wearing clothes once and not washing with extra detergent.

13.3.5 Encouraging Cutting-Edge Technology

Proposing to consumers new lifestyles of clothes washing that encourage advanced technologies is critical. For example, encouraging the use of laundry services where experts use automated washing machines with special sensors that can detect the level of detergent and water is required.

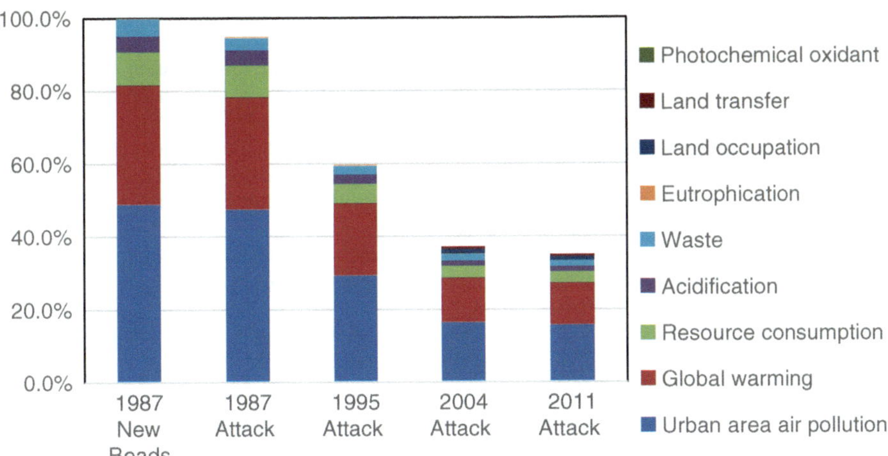

Fig. 13.3 Evaluation of the (area-specific) environmental impacts per clothes washing (detergent: New beads, Attack manufactured by Kao Corporation)

13.3.6 Understanding Diverse Washing Technologies and Culture

Because washing has developed throughout world history, there are a wide variety of washing methods and norms. It is necessary to understand the diversity of washing technologies and cultural practices. Different washing techniques might hint at new technologies that can help enable the future of washing. The diversity of washing methods has been developed with local values, and it is important to consider those values in the future of washing. If those values are not respected, there could be distortions and the future of washing may not be achieved.

In order to create the future of washing, a wide range of research and fieldwork needs to be conducted, including: assessing the environmental and social impacts of washing; developing new sustainable washing technologies such as detergents and washing machines; and taking into consideration behavioral science including research about the history and culture of washing. Moreover, key actors, such as government, academia, and the private sector need to come together to discuss the various aspects of washing. In 2018, Future Earth [7], The University of Tokyo Institute for Future Initiatives [8], and Kao Corporation [9] established the Future of Washing Initiative [10] in order to lead a discussion and make suggestions about the future of washing.

The future of washing is essential for our lives and represents an important theme for achieving a sustainable society. It is critical to gather diverse knowledge and ideas from experts and consumers in order to begin discussions about the future of washing.

References

1. Japan Soap and Detergent Association. https://jsda.org/w/02_anzen/index.html
2. Ministry of the Environment Global Forest Conservation. http://www.env.go.jp/nature/shinrin/fpp/certification/index3-2.html
3. Japan Chemical Industry Association: Innovations in greenhouse gas reductions (March 2014). https://www.nikkakyo.org/sites/default/files/cLCA_3_factsheet2014-3-18_0.pdf
4. Ministry of Land, Infrastructure and Transport. https://www.mlit.go.jp/mizukokudo/mizsei/mizukokudo_mizsei_tk2_000014.html
5. Bureau of Waterworks Tokyo Metropolitan Government. https://www.waterworks.metro.tokyo.jp/kurashi/shiyou/jouzu.html, https://www.mlit.go.jp/mizukokudo/mizsei/mizukokudo_mizsei_tk2_000014.html
6. Ohtawa Y, Miura H, Itsubo N. The change of environmental impact of compact powder laundry detergents by LIME2 analysis. J Lifecycle Assess, Japan. 2015;11:300.
7. Future Earth. https://futureearth.org/
8. The University of Tokyo Institute for Future Initiatives. https://ifi.u-tokyo.ac.jp/en/
9. Kao corporation. https://www.kao.com/global/en/
10. Future of Washing Initiative. https://www.futurewashing.org/

Chapter 14
Initiatives on Responsible Consumption: Oguni Town

Emi Mori

Abstract Oguni Town is a small town located in the western part of Mount Aso in Kyushu, which was selected as an "Eco-Model City" by the Japanese government in 2013. After being selected as an SDGs Future City, Oguni Town formulated the "SDGs Future City Plan" and the "SDGs Future City Action Plan." The goal is to realize a sustainable town by creating a recycling-oriented society and industries utilizing geothermal heat and forest resources. This chapter describes the past progress of this project.

Keywords Eco-Model City · Geothermal heat and forest resource · Low-carbon society · Oguni Town · SDGs

14.1 Introduction

In September 2015, the United Nations Sustainable Development Summit was held at the United Nations Headquarters in New York City, and the outcome document on the 2030 agenda for sustainable development was unanimously adopted. The bottom line of the 2030 agenda is the realization of SDGs. In May 2016, the Japanese Cabinet approved the establishment of the Headquarters for the Promotion of SDGs headed by the Prime Minister, and the Japanese SDGs Promotion is in full swing. The promotion of the SDGs in local governments was accelerated in the fiscal year 2017 when the author was in charge of the regional development in the Cabinet

E. Mori (✉)
Oguni Town Hall, General Affairs Division, Kumamoto, Japan
e-mail: e_mori@town.kumamoto-oguni.lg.jp

T. Nakajima et al. (eds.), *Overcoming Environmental Risks to Achieve Sustainable Development Goals*, Current Topics in Environmental Health and Preventive Medicine, https://doi.org/10.1007/978-981-16-6249-2_14

Office on secondment. In the same year, based on the summary of the[1] "Advisory Panel for the Promotion of Local Government SDGs" on the basic concept of promoting local government SDGs, we, the Secretariat of Local Government Development Promotion (Local Government Development Promotion Office) of the Cabinet Office, decided to proceed with the selection of "SDG Future Cities."[2]

This chapter describes how Oguni Town, a small farming and mountainous village in Kumamoto Prefecture located in Kyusyu Island of Japan with a declining population and serious regional challenges, envisions its future in 2030, taking the opportunity of being selected as an SDG future city, and how it aims to become a sustainable town by utilizing geothermal energy to achieve this goal.

14.2 From Environmental Model City to SDGs Future City

Oguni Town is located in the western part of the Kokonoe Mountain Range, north of Mount Aso, and has underground steam and hot water (hereinafter referred to as "geothermal resources"). It is a depopulated mid-mountainous area with a population of about 7200 people. In FY2013, the town was selected as an "Environmental Model City" by the government to realize a low-carbon society, and has been promoting "urban development focusing on the environment." In FY2018, the town was selected as an "SDGs City of the Future" for its proposal on "urban development using regional geothermal energy and forests."

After the selection of the SDGs Future City, Oguni Town formulated the "SDGs Future City Plan" and "SDGs Future City Action Plan." The town in 2030 should be "a sustainable city that creates a recycling society and industry by utilizing local resources," and we[3] are working toward achieving it.

14.3 Responsibility to Use

The core of Oguni Town's plan to become an SDG future city is a geothermal power project (hereinafter referred to as "geothermal power generation"). However, geothermal power generation in our town has an unfortunate history as a major national project had been withdrawn from our town in 2002 due to the lack of understanding

[1] Cabinet Secretariat/Cabinet Office General Website (https://www.kantei.go.jp/jp/singi/tiiki/kankyo/kaigi/sdgs.html), accessed Jan 15, 2021.

[2] Hakaru Tamura (2019): "Promoting SDGs that Create Regional Revitalization", The first International Forum on SDGs for Regional Revitalization (https://future-city.go.jp/data/pdf/event/2019/20190213_Hakaru_Tamura_en.pdf), accessed April 25, 2021.

[3] Oguni Town website (https://www.town.kumamoto-oguni.lg.jp), accessed Jan. 15, 2021. Emi Mori (2020): "Initiatives of a Japanese SDGs Future City "Oguni Town"—On Utilization of Geothermal Resources –", Journal of Groundwater Hydrology, 62(1), p 33–42.

among residents. Opponents of geothermal power generation were concerned that the increased "use" of geothermal resources through development would lead to the decline of hot springs and cause micro-earthquakes and strongly argued that we should avoid large-scale development and keep the use of geothermal resources as an integral part of our lives such as running of hot springs in homes and inns, the using of steam for heating, etc.

In 2014, 12 years after the national government withdrew the project, a group of pro-geothermal power generation residents surprisingly established a joint venture company to operate a geothermal power plant in the same area.[4] At present, all ex-opposition residents have also invested in the company and are participating in the geothermal power generation project. Two factors helped them to reach this point without government intervention: first, the attention to renewable energies has become a tailwind after the Great East Japan earthquake. In particular, the accident at the Tokyo Electric Power Company Fukushima Daiichi Nuclear Power Plant, which occurred as a result of the tsunami triggered by the March 11, 2011 earthquake off the Pacific coast of Tohoku, resulted in the meltdown of the nuclear reactor and the resulting release of radioactive materials, leading to the evacuation of nearby residents. It also revealed many issues with nuclear power generation, such as the limited capacity of power supply in Japan. Second, most people that had opposed the project now retired, and their successors who had felt the limitations of their business began to realize the necessity of promoting geothermal power generation.

In recent years, many geothermal power producers, especially those from outside of the town, try to enter the power production business in Oguni. Therefore, in 2016, Oguni Town enacted the "Ordinance on Utilization of Geothermal Resources in Oguni Town" to promote appropriate and sustainable utilization of geothermal resources, contributing to the progress of the local economy and proper management of the utilization of geothermal resources.

Geothermal resources are valuable assets unique to the town. Its effective and efficient use will lead to the creation of a sustainable city. In Japan, where energy resources are scarce, renewable energies based on geothermal resources are expected to play an important role as a base-load power source, because of not only their low generation costs but also their ability to generate power stably. However, if geothermal resources are not used properly or overused, it may cause not only a decrease in the overall underground resources but also jeopardize the livelihood of the local people and the sustainability of the town. Therefore, careful handling of geothermal resources is also required. In addition, it is important to stress the responsibility of developers who use geothermal resources for power generation projects 14, as well as the responsibility of people who use the power, to assure the residents' livelihood security.

[4] RENEWABLE ENERGY INSTITUTE Website (https://www.renewable-ei.org/column_r/REapplication_20170725.php), accessed May. 3, 2021.

14.4 On Behalf of the Knot

Geothermal power generation not only leads to the realization of local energy production for local consumption and the construction of a disaster-resistant and environmentally friendly town but also increases the GDP per capita.[5] It is our mission as a government to create a recycling-conscious society and industries that make use of local resources efficiently. The public and private sectors must work together to protect the natural environment, including geothermal resources inherited from our predecessors, while communicating constructively with residents for appropriate geothermal power development.

[5]According to "The Kumamoto Earthquake and Local Economies" (Regional Economic Research Institute, February 2019), the GDP per capita in Oguni Town, which is currently 2.37 million yen, is estimated to be 3.5 million yen (+1.13 million yen) in 2030 due to geothermal power generation and other initiatives.

Chapter 15
Lessons of Court Decisions on Minamata Disease and Future Actions

Tadashi Otsuka

Abstract Two judgments of Minamata Disease, the judgment of Kumamoto District Court, March 20, 1973 and the judgment of Supreme Court, October 15, 2004 still give us many lessons. I would like to point out two points.

Firstly, the case of Minamata Disease and these judgments are linked to the contemporary environmental policy to minimize the utilization and release of hazardous chemical substances. Minamata Convention on Mercury was adopted with this aim. It is related to SDGs, Target 6.3 aims and 3.9.

Secondly, both judgments dealt with the contemporary precautionary principle, the issue of how the government and companies should respond to the scientific uncertainty problems. Up to now, we have had quite a few scientifically uncertain issues. In this situation, on the one hand, we should seek scientific evidence and on the other hand, we should make rational choices in society. The experience of Minamata Disease also seems to teach us many lessons for the application of the Precautionary Principle to environmental burden.

Keywords Minamata Disease · State responsibility · Precautionary principle Negligence · SDGs

15.1 Introduction

Minamata Disease is the most serious pollution case in Japan. As for the judgments about pollution in Japan, I would like to deal with two decisions regarding the Kumamoto Minamata Disease case.

T. Otsuka (✉)
Waseda Law School, Tokyo, Japan
e-mail: totsuka@waseda.jp

T. Nakajima et al. (eds.), *Overcoming Environmental Risks to Achieve Sustainable Development Goals*, Current Topics in Environmental Health and Preventive Medicine, https://doi.org/10.1007/978-981-16-6249-2_15

There has been an extensive scientific discussion about the notion of Minamata Disease. I would like to entrust the explanation of the notion of Minamata disease to the author of another chapter in this book.

First, I will give a brief history of Minamata Disease cases. Next, I will introduce two decisions regarding the Kumamoto Minamata Disease case. Finally, I would like to consider some lessons learned from them for examining future regulations of chemicals.

15.2 Brief History of Minamata Disease Cases

Minamata Disease cases include both Kumamoto Minamata Disease cases and Niigata Minamata Disease cases. Kumamoto Minamata Disease cases occurred in the area surrounding Minamata City in Kumamoto Prefecture from about 1953 to the early 1960s. The methyl mercury contained in the water discharged from the acetaldehyde manufacturing facility of Chisso Minamata factory accumulated in the fish in Minamata Bay and the area surrounding Minamata Bay. The area's residents, ingesting the fish repeatedly, contracted Minamata Disease. In 1956, this disease was officially discovered when the hospital attached to Chisso Co. reported it to the Minamata Public Health Center. Soon the Medical Faculty of Kumamoto University suspected this disease was caused by the water discharged from the Chisso factory. However, in 1968 the government finally officially announced that the cause of Kumamoto Minamata Disease was methyl mercury compounds discharged by Chisso Co (For legal treatises of Kumamoto Minamata Disease litigations in 1990s [1–3]). Meanwhile, in 1964 and 1965, the same methyl mercury toxicosis occurred in the basin of the Agano River in Niigata Prefecture. The government officially recognized that Niigata Minamata Disease was caused by the water containing methyl mercury compounds discharged from Kase factory of Showa-Denko Company (For the first Niigata Minamata Disease litigation, Niigata District Court issued a decision on Sept. 29th, 1971 [4]).

I would like to show a brief chronology of the Kumamoto Minamata Disease case in Table 15.1.

15.3 Two Decisions Regarding the Compensation in the Minamata Disease Cases

I would like to take up two decisions regarding the compensation in the Minamata Disease cases: firstly, a compensation litigation filed against the company that directly caused this disease; secondly, a State responsibility compensation litigation because of the failure of the central and prefectural governments to exercise their authority.

Table 15.1 Chronology of Kumamoto Minamata Disease case

Year	Chronology of Events
From 1950	Shellfish and fish died in Minamata Bay and in the coastal areas, cats went mad and died of what is called "dancing cat fever".
1953	In Minamata City, a girl developed a disease. Later she was identified as the first patient of Minamata disease.
1954	A man who lived in Minamata City complained of **concentric contraction of the visual field and was admitted to the Chisso Affiliated Hospital.**
November 1956	Kumamoto University research group announced in an interim report that this disease is a heavy metal poisoning and it is mainly due to the intake of seafood
October 1957	Health Science Research Group announced that the causal substances of Minamata disease are manganese, selenium, and thallium. It was suspected that they were released from Chisso Corporation.
March 1959	Water Quality Conservation Act and Factory Wastewater Regulation Act came into force.
July 1959	Kumamoto University research group announced officially that this disease is a neurological disorder caused by the intake of poisoned seafood and as for the poison, the focus centered on mercury.
August 1959	Prof. Kiyoura published in a newspaper his opinion that the methyl mercury theory was doubtful and it should be dealt with prudently.
October 6, 1959	The subcommittee of Minamata Food Poisoning (a special committee of Food Sanitation and Investigation Council, an advisory body of the Health Minister) announced an interim report that Minamata disease is very similar to methyl mercury poisoning and that mercury is the most suspicious causal substance.
November 12, 1959	Based on the interim report, Food Sanitation and Investigation Council concluded that the main cause of Minamata disease is a kind of methyl mercury and it submitted the report to that effect to the Health Minister.

1. The first decision [5] is regarding the First Compensation Litigation of Kumamoto Minamata Disease that was filed by the patients of Minamata Disease against Chisso Co. Kumamoto District Court recognized the negligence of Chisso and applied Sec. 709 of Civil Code, the general provision regarding torts.

   According to this decision, "A chemical company which discharges wastewater from its factory has a high-level due diligence to confirm the safety of its wastewater by fully investigating with the best knowledge and techniques whether or not the wastewater includes a hazardous substance. The chemical company also has a high-level due diligence to take the maximum required preventive measures, especially to prevent danger to the life and health of the local residents, by immediately stopping the operation of the factory in the unlikely event that the wastewater is found to be hazardous or its safety becomes doubtful. While it is the companies that discharge wastewater, local residents have no means of knowing what products those companies make, how such products are produced or what kind of wastewater is discharged from their factories. Therefore, naturally, a chemical company has the unilateral obligation to insure the safety to the life and health of the residents." "It is clear that" the plaintiffs "contracted Minamata Disease during the period between 1956 and 1961."

This decision is understood to mean that chemical companies have an obligation to investigate all substances they discharge from their factories, not only in the situation of scientific uncertainty of 1956, where scientific data were found but the causal substance was not identified but also in the situation where scientific data were not found. Such understanding of this decision is proved because it identified the negligence of Chisso regarding the plaintiffs who contracted Minamata Disease in 1953. Even today this decision has significant implications.

2. The second decision [6] is regarding the Supreme Court decision about the Kansai Minamata Disease case where the plaintiffs demanded State compensation.

The Supreme Court ruled that the failure of the Minister of Trade and Industry to exercise his authority to designate Minamata Bay as the designated area based on the Public Water Area Quality Conservation Act and the Factory Discharge Regulation Act and to set a water quality standard had been significantly unreasonable since January 1960, and it had become illegal to apply Sec. 1 of the State Redress Act. It ruled that the Minister of Trade and Industry could have exercised his authority by the end of December 1959, because at the end of November 1959, (1) the State recognized that there was a considerable number of deaths resulting from Minamata Disease and (2) the State could recognize with high probability that the causal substance of Minamata Disease was a kind of methyl mercury and its source was a facility of Chisso Minamata factory.

It also ruled that the failure of Kumamoto Prefecture to exercise its authority to direct Chisso to set any necessary equipment to remove hazardous substances in the wastewater from its factory was illegal.

This decision has two characteristics.

Firstly, this is the first supreme court decision to admit State compensation because of the failure of the central and prefectural governments to exercise their authority regarding pollution. Before this decision was made, it was considered very difficult to demand compensation from the State because of the failure of the government to exercise its authority regarding pollution. However, owing to this and other decisions, the State recognized that the inaction of the government to exercise its authority at an appropriate time and in an appropriate manner could be illegal. The fact that this decision ruled that the failure of the State to exercise its authority had become illegal since January 1960 was a new lesson to the State.

Secondly, there is room to see that this decision and the original decision [7], overcame scientific uncertainty in factfinding. The first instance, Osaka District Court, July 11, 1994 ruled that in 1950s methyl mercury could not be measured and mercury could be measured only as the total mercury, that regulating by total mercury became overregulation and that the discharge from Chisso factory would probably be less than the detection limit value. However, Osaka Court of Appeal ruled that the argument of overregulation could not be adopted considering the necessity to prevent the significant damage of Minamata Disease. Regarding the detection limit, it seemed to take no notice of the reproducibility of data considering "the necessity to prevent the further occurrence of damage as soon as possible." The Supreme Court did not mention the issue of overregulation and determined that "the quantitative analysis was possible."

15.4 Lesson from the Two Judgments

Thus, the two judgements regarding Minamata Disease, on the one hand, imposed on a chemical company a high-level due diligence to investigate the substances discharged from its factory, and on the other hand, showed that the central and local governments have the obligation to exercise their authorities at an appropriate time and in an appropriate manner and that the failure to exercise their authorities may be illegal.

Both judgments have very significant implications and will become the basis of activities of companies and administrations regarding environmental issues. The above-mentioned obligations imposed on companies and administrations are also very important for scientists related to companies and administrations.

The lessons of the two judgments are not limited to this. I would like to point out two points.

Firstly, the case of Minamata Disease and these judgments are linked to the contemporary environmental policy to minimize the utilization and release of hazardous chemical substances. Minamata Convention on Mercury was adopted with this aim. It is related to SDGs, Target 6.3 aims and 3.9.

It is our mission to link the lessons of Japanese pollution cases, especially Minamata Disease cases, to the global minimization of the utilization and release of hazardous chemical substances.

Secondly, both judgments dealt with the contemporary precautionary principle, the issue of how the government and companies should respond to the scientific uncertainty problems.

After the occurrence of Minamata Disease, Precautionary Principle expanded in Europe starting with West Germany at that time and has been adopted in international documents. Though there are so many documents dealing with this principle, the 1992 Rio Declaration and 2000 EU communication paper are the most important.

Up to now, we have had quite a few scientifically uncertain issues. For example, the risks of health damage due to marine pollution by plastics or the nonthermal effect of electromagnetic waves are still scientifically uncertain. In this situation, on the one hand, we should seek scientific evidence and on the other hand, we should make rational choices in society. The experience of Minamata Disease also seems to teach us many lessons for the application of the Precautionary Principle to environmental burden.

Thirdly, I will compare Minamata Disease cases and Fukushima Daiichi nuclear power plant accident case. Fukushima case is very different from Minamata cases in that it is not a nuisance case. It is a pollution case resulting from an accident and is not a continuous pollution case. It is important to understand this difference. The legal lesson from Minamata Disease cannot be directly applied to Fukushima case.

I have two comments regarding nuclear accident cases. (1) Nuclear power plant operators are subject to the no-fault liability principle under the Nuclear Damage Compensation Act. This is very different from ordinary pollution cases because ordinary factory operators including Chisso are subject to the fault liability principle under the Civil Code. As for nuclear power plant operators, we do not have to consider whether they were at fault for causing the damage to victims. We do not have to take into account the Precautionary Principle here. (2) Regarding the State, it is subject to the fault liability principle. Lower court cases are divided into two types: those that admitted the fault of the State and those that did not admit the fault of the State. The way of understanding the concept of fault, i.e., the way of dealing with the Precautionary Principle, reflected the difference between the two types of decisions. This is because Fukushima accident resulted from the Great East Japan Earthquake and the prediction of the possibility of the great earthquake occurrence around Fukushima before the accident was the most important issue. Whether and to what extend the great earthquake would hit Fukushima were scientifically uncertain before the earthquake. The Supreme Court will make a decision in the near future.

Fourthly, COVID-19 is still spreading around the world and many past actions of the Chinese and Japanese governments and others may have influenced the current situation of infections in Japan. Though we can assess these actions from the viewpoint of the Precautionary Principle, it is very unclear whether some of the actions can be legally connected to the liability of the aforementioned.

References

1. Sadao Tomigashi, Minamata disease cases and the law (1995, Sekifusha).
2. Otsuka T. Comprehensive deliberation of Minamata disease decisions (1)–(5). Jurist. 1996;1088:21–9. vol. 1090, pp. 81–89; vol. 1093, pp. 101–105; vol. 1094, pp. 109–112; vol. 1097, pp. 109–112
3. Otsuka T. The significance and task of the supreme court decision of Kansai Minamata disease litigation. Hanrei Times. 2006;57(3):91–9.
4. Niigata District Court, Sept. 29th, 1971, Kakyu Minji Hanreisyu, vol. 22–9=10, Supplementary p. 1.
5. Kumamoto District Court, March 20th, 1973, Hanrei-jiho, vol. 697, p. 15.
6. Supreme Court, Oct. 15th, 2004, Saikosaibansho Minji Hanreisyu, vol.58–7, p. 1802.
7. Osaka Court of Appeal, April, 27th, 2001, Hanrei-jiho, vol. 1761, p. 3.

Chapter 16
An Initiative of an Environmental Model City: Featuring Sustainable and Healthy Cities

Keiko Nakamura

Abstract People of the city of Minamata, Japan have worked on creating an environmental model city since the early 1990s because of their desire to "do something about Minamata." Government recognition of Minamata as an environmental model city followed in 2008. Along with promoting relief measures for Minamata disease victims and reconciliation among the citizens, initiatives to protect the environment, respect people's health and welfare, and cherish natural ecosystems have emerged. Citizens themselves led projects to promote environmental measures, created local bases for activities, and planned and implemented city projects. Community-led recycling projects, the city's self-declaration of compliance to international standards for environmental management systems corresponding to audits by citizens, promotion of zero-waste measures, and other programs encouraged citizens to practice eco-friendly actions in their daily lives. Minamata's experience toward becoming an environmental model city will accelerate global initiatives for Sustainable Development Goals (SDGs) by passing shared values for sustainability and health from generation to generation.

Keywords Minamata disease · Environmental model city · Zero waste Sustainable Development Goals (SDGs) · Healthy cities · Value-based urban planning

K. Nakamura (✉)
Department of Global Health Entrepreneurship, Tokyo Medical and Dental University, Tokyo, Japan
e-mail: nakamura.ith@tmd.ac.jp

T. Nakajima et al. (eds.), *Overcoming Environmental Risks to Achieve Sustainable Development Goals*, Current Topics in Environmental Health and Preventive Medicine, https://doi.org/10.1007/978-981-16-6249-2_16

16.1 Introduction

Areas that have experienced environmental pollution, experience not only excessive burdens on human health and the environment but also various burdens on the entire community. For example, in the case of Minamata disease, pollution of the sea by methylmercury not only caused enormous damage to the health of the ecosystem and population but also caused tangible and intangible conflicts between affected patients and other people in the area. In addition, the community of Minamata as a whole experienced alienation and discrimination from other communities. Through these experiences, Minamata City, Kumamoto Prefecture, Japan, found a clue to rebuild the community and worked on creating an Environmental Model City ahead of the rest of the country. While incorporating innovative initiatives and programs into people's daily lives, for the last 30 years, the city has made continuous efforts to feature environment-oriented community development. Based on the progress in creating an environmental model city in the area where Minamata disease was experienced, implications for other cities to achieve Sustainable Development Goals (SDGs) will be discussed in this chapter.

16.2 Creation of an "Environmental Model City" Based on the Lessons Learned from Minamata Disease

A patient with neurological symptoms officially reported in 1956 was later known as the first reported case of Minamata disease. In 1968, 12 years since the first case, the central government officially presented a view on Minamata disease as an issue caused by methylmercury discharged from the Chisso Minamata factory. Patients diagnosed with Minamata disease and other local people were at the mercy of the problem of Minamata disease. While early relief for Minamata disease victims was prioritized from the beginning, Kumamoto Prefecture also started a pollution control project in 1977. A large-scale sediment disposal study was conducted in Minamata Bay. Sediment, dredged from the remaining contaminated area in the bay, was discharged into the reclamation area in the inner area of the bay with careful measures to prevent additional pollution. Removal of mercury-contaminated sediment in the bay was confirmed according to government standards, and the project was successfully completed by 1990 [1].

Rebuilding a community in Minamata was a process that started from a difficult situation, as the city's economic, cultural, and social foundations were devastated at the time. In addition to the tremendous health damage caused by methylmercury pollution, there were serious social divisions among citizens. The society became a swirling center of conflict, malicious slander, and derogatory behavior. Despite those struggles, it is believed that the citizens' desire to "do something about Minamata" led to the movement of regenerating Minamata, beginning the first half of 1990 [2].

With the completion of the Minamata Bay pollution control project by the prefectural government, the movement to regenerate the community began. The

Minamata Disease Municipal Museum and the Kumamoto Environmental Center were founded at this time, and a memorial service for Minamata disease victims was organized for the first time. A series of local public meetings to discuss issues related to Minamata disease was followed by the "Environmental Creation Minamata" promotion project, a joint work of Kumamoto Prefecture and Minamata City. The Minamata Promotion Office supported by the prefectural government created structured and organized events in which citizens could participate, such as the opening of the Eco Park Minamata, a world bamboo forest park that collects bamboo from around the world, and the "Environmental Creation Minamata '92" events. Through these efforts, the momentum for environmental protection began to grow, and the initial step was taken toward becoming an environmental model city that emphasizes environmental protection [2].

The Minamata City Council declared "Development of City that Treasure Environment, Health, and Social Welfare" in June 1992 (Box 16.1) [3]. For the first time in Japan, Minamata City declared "Creating an Environmental Model City" in November 1992 (Box 16.2) [4]. These declarations emphasized the importance of environmental protection, health and welfare in community development and reflected a strong determination to prevent the re-occurrence of tragic incidents by sharing information with the public about the suffering and the lessons learned from Minamata disease [2].

Box 16.1 The Declaration of the Development of City that Treasure Environment, Health, and Social Welfare [3]

The City of Minamata has been experiencing Minamata disease, an industrial pollution with degree of destruction so strong that damages the health of man and other living creatures, as well as the ecosystem, which was unprecedented in the history of mankind. The disease broke out when Japan placed the highest priority on production as we struggled to achieve a rapid economic growth during Japan's postwar restoration. The Citizens of Minamata are trying to overcome these unprecedented damages and their immeasurable affect by getting together and striving to achieve their goal by making use of their valuable lessons. The people of Minamata City aim to developing their city into a place that treasures the environment, and where the nature, the people, and the industries can exist in harmony.

In the midst of increasing understanding concerning the destruction of the environment which is the biggest part of the world crisis these days, informing the people throughout the world about the lessons that we have experienced with Minamata disease can be a warning bell to those who are about to follow the same footstep of Minamata. It is our understanding that by expressing our concerns for the environment in such way, we, although were once known as the source of public pollution, can fulfill the greatest role of contributing to the world's environmental issues and problems in a positive way.

Because we must never forget the sacrifice caused by the disease, whereas we must never let the same disaster repeat again, whereas: we must continue the relief of those who are still suffering from the disease, whereas we must preserve the dignity of life as we learned from our past experience of Minamata disease incident, the citizens of Minamata hereby declare the promotion of the city development with the aim to create a place that preserves environment and health, and values welfare.

June 25, 1992
Minamata City Council

Box 16.2 Declaration of Creating an Environmental Model City [4]
The history and climate of Minamata City pose questions about what human beings should do to conform and coexist with nature. From time immemorial, people have lived near fields and mountains, always in places near water sources such as rivers and the sea which have provided them with ample natural resources.

However, the outbreak of Minamata disease, which is said to be one of the world's biggest and worst cases of industrial pollution resulting from a productivity-first principle, has severely affected human health and the environment in the vicinity of Minamata city. This highlights the dangers of industrial development. It has caused death not only to citizens but to many other living things, brought about psychological anguish, and endangered the very existence of local communities. As a matter of fact, Minamata's citizens have suffered for 36 years.

Learning from this tragic experience, we in the City of Minamata are determined to construct a model environmental city that will respect natural ecosystems. We are also determined to disseminate the lessons we have learned from Minamata disease all over the world:

1. Passing the lessons learned from Minamata disease down to the next generation
2. Promoting relief measures for Minamata disease victims and reconciliation among the citizens
3. Encouraging changes to industrial activities that will not endanger the people and all other living things in cyclic natural ecosystems
4. Protecting such fundamentals for life as the sea, rivers, and mountains and handing them down in good condition to the next generation
5. Reviewing the structure of present-day civilization and propose a new social system that will be based on the frugal use of limited natural resources

We hereby declare that in view of the "Environment Creation, and Minamata '92" events taking place in Minamata, we in the City of Minamata reaffirm our resolution to implement the undertakings we have stated above.

November 14, 1992
Minamata City

16.3 A Collective Vision of the Future and the Community's Continuous Effort

Minamata's vision of creating an environmental model city was shared among the public, and further progress was made, including enhancing relief measures for Minamata disease victims, advancing public understanding of Minamata disease issues, promoting community revitalization, and developing a park that promotes values of health and environmental consideration at the Minamata Bay reclaimed land. In addition, "*Moyai-Naoshi*" (reconnecting knots in communities) was promoted with the aim for citizens to work together regardless of their individual perspectives. "*Moyai-Naoshi*" is a symbolic word that represents the strenuous effort to tackle the Minamata disease, and also implies the restoration of relationships within the community. This is especially important because the outbreak of Minamata disease instigated conflicts, slander, and accusations, which resulted in serious problems and severed relationships among community members. Through the "*Moyai-Naoshi*," people strove to achieve their goals and make decisions by having constructive discussions based on recognition of each other's opinions and respect for individual situations [2].

Minamata City worked on the acquisition to satisfy ISO 14001, the international standards for environmental management systems, and acquired its recognition in 1999. This led to the establishment of the city's own system to self-declare ISO 14001 corresponding to environmental audits by citizens [5].

In addition, the city developed its own standards for different settings and called them ISO for schools, ISO for households, ISO for nurseries/kindergartens, and ISO for inn/hotels. This enabled each setting to address specific eco-friendly measures and promote their efforts. These activities served as community-based environmental learning systems for citizens to acquire knowledge on how to practice eco-friendly actions in their real lives.

Moreover, citizens themselves led the projects of promoting environmental measures and created local bases for activities through citizens' organizations such as the Association of Contemplations and Actions for the Environment of Minamata (a collaborative body of 17 citizen groups), as well as the "*Yoro-Kai*" (meaning "let's get together," and formed in each of the 26 administrative districts of Minamata City). Further, systems for citizens' participation in the planning and implementation of the city's projects were developed, and citizens participated as members of the Basic Plan Conceptualizing Committee for City Development [5].

Since 1993, Minamata City has implemented a system of garbage classification and recycling household waste using a detailed classification system. At that time, garbage was divided into 20 groups within five categories. The city became well known as a pioneer of efforts to sort household garbage thoroughly and resource recycling. A total of 330 garbage collection stations were set up across the city and citizens worked together to sort the garbage. This system has become a national and international benchmark for community-led recycling projects, as a model of family and community engagement through learning in daily life [5].

Multi-stakeholder involvement is another characteristic of a community's efforts. Citizens, industries, and local governments collaborated to plan and implement programs and encourage practical advancement. Projects of reusing and recycling glass bottles suggested enjoyable lifestyles for citizens, as well as expanded innovative use of recycled glass and their markets. Projects of recycling plastic compounds gained their value in the market because of their high quality, supported by the awareness of citizens through the separation of household garbage. Citizens' daily thorough practices toward recycling, application of innovative technologies by industries, and support by the local government through subsidies, relaxation of regulations, promotional activities, city-wide movement, and others also promoted recycling practices [6]. Learning from practical examples of programs and activities across stakeholders contributed to the accumulation of concrete examples that would realize Minamata's vision for the future.

16.4 Shared Values Among the Community and Passing These onto the Next Generation

Recently, many urban development projects have been taking into account the environment and the health of citizens. The concept of value-based urban planning has been advocated [7].

Since the beginning of the movement of Minamata's regeneration in the 1990s, citizens have strived to cherish common principles such as preserving the environment, health, welfare, the natural cycle of the ecosystem, recycling of resources, harmonization of the environment and the economy, sustainable community and citizens' engagement. When the notion that environment-friendly concepts would hinder earnings and economic productivity was still prevalent, the citizens of Minamata took the initiative to create an environmental model city that valued health and the environment and engaged in daily efforts to achieve this goal [2].

During the late 2000s, Minamata City was nationally and internationally recognized for its efforts to protect the environment, and the Cabinet Office of Japan officially recognized Minamata City as an environmental model city in 2008. Since then, the Basic Environmental Ordinance has proclaimed building a low-carbon society. Moreover, various projects linked to concepts such as the utilization of limited resources and energy, as well as the development of the local economy by considering the unique characteristics of the areas, were planned [8]. Furthermore, the completion of Eco Park Minamata with the theme of environment and health and the Minamata Declaration for Creating a Zero-Waste City (Box 16.3) in 2009 [9], demonstrated concrete examples of valuing the environment and health. Such programs and activities have accumulated since the 1990s.

Box 16.3 Minamata Declaration for Creating a Zero Waste City [9]

We, the citizens of Minamata, shall examine our lifestyle, treasure the blessings of nature and exert effort to build a life and system that makes optimal use of our limited resources and energy.

We shall develop an urban system by 2026 that will not depend on incineration and landfill for waste disposal while maintaining the precious nature of our home and protecting the lives and health of all living things.

We shall have pride in our endeavors and the efforts we put into achieving our goal since Environmental Model City Declaration in 1992 and move forward with Zero Waste measures utilizing our past experience and results.

We shall share information necessary to move forward with Zero Waste projects among citizens, businesses and administration, set up a venue where discussions can be held on a continuous basis, and cooperate to review goals, actions and achievements.

We shall join hands with all people and municipalities in Japan and the world with the common ambition, aim to solve problems together, and increase the number of people joining in Zero Waste projects in order to bring a better environment to Japan and the world.

November 22, 2009
Minamata City, Kumamoto

Minamata has been involved in international environmental projects since the United Nations Conference on Environment and Development (Earth Summit) in 1992. Thus, it was natural for Minamata City to take action toward achieving the SDGs adopted by UN member states in 2015. In 2018, Minamata City was designated as an "SDGs Future City" by the Cabinet Office of Japan.

Three generations have passed since the outbreak of Minamata disease. The first generation had already wrestled with various obstacles in the 1970s as adults, the second generation spent their childhood in the 1980s and grew up with difficulties related to Minamata disease faced by the community, and the third generation leaned about Minamata disease as a history of the community and grew up with eco-friendly lifestyles. In the near future, the third generation will take over the roles of leading the city toward a sustainable and healthy city.

Yet, there are predictions that the city's population will continue to decline in the coming years. In 1995, Minamata City's population was approximately 50,000, which was reduced to approximately 25,000 by 2015. By 2030, the population is expected to be less than 20,000. At the same time, the city is faced with the challenge of fostering industries for a sustainable local economy. Hence, it is important to pass down the lessons learned from Minamata disease to the younger generations who did not witness the issues directly. A continuous learning system should be established to encourage learners to make new contributions to the future. These efforts will enable the community to share values with people not only in Minamata but also outside of Minamata and who did not directly experience the issues of Minamata disease.

Minamata children are deepening their understanding of the importance of respecting lives, caring about health and welfare, and protecting the environment by studying Minamata disease, listening to the various life stories surrounding Minamata disease from local adults, and engaging in recycling activities in their daily lives. They also have rigorous environmental education, including learning eco-friendly lifestyles through ISO activities at their schools. Additionally, they have heard about and/or witnessed the *"Moyai-Naoshi"* project. In the near future, when this generation becomes the center of the city's development, it will be worthwhile to see how they implement environmental and health values with compassion, as these are the foundational concepts of Minamata's environmental model city project. As the methods of sharing values diversify, our responsibility across communities is to ponder effective approaches to share these values and ensure that they are passed onto the next generations.

16.5 Conclusion: Achieving Sustainable Development Goals

In order to achieve the goals of the 2030 Agenda for Sustainable Development (SDGs), the initiatives of national governments and international organizations, as well as the actions of local governments and communities, are emphasized. There are 17 goals within SDGs, which include climate action, quality education, good health and well-being, clean water and sanitation, industry/innovation and infrastructure, affordable and clean energy, decent work and economic growth, and sustainable cities and communities. Since many of these goals are mutually associated, a cross-disciplinary approach is encouraged.

For most of the citizens of Minamata, recognizing the interrelationship of the SDGs' 17 goals is self-evident because the citizens have been engaging in these actions in their daily activities over 30 years. The goals had already been stated in the "Declaration of Creating an Environmental Model City" since the beginning of the 1990s. Examples of specific actions and projects were continuously conducted under the policies of Zero-Waste city and action plans for an environmental model city.

While facing several challenges related to its declining population, along with the low birth rate and aging population, Minamata's long-term experiences of tackling issues of environmental pollution and its influence on communities, enabled the city to share their creative efforts as a model for other communities. It is highly anticipated that the city will contribute to accelerating the global initiatives for SDGs by passing shared values for sustainability and health from generation to generation and by expanding opportunities to share their experiences in Japan with international communities.

Acknowledgments Sincere appreciation is extended to the citizens of Minamata for their collective work toward an environmental model city. Appreciation is also extended to Minamata City for documentation of the process toward an environmental model city and for sharing this information. The author appreciates Dr. Noriyuki Hachiya and Ms. Rie Harada of the National Institute for Minamata Disease and Minamata Disease Information Center for their comments and information shared.

References

1. Hachiya N. History and community of Minamata disease: Minamata under the modernization and high economic growth of Japan. Modern Kumamoto (Kumamoto Modern History Study Group) No. 39, pp. 49–70, 2017. (in Japanese).
2. Yoshii M. "Janaka Shaba" creation of new Minamata. Fujiwara Shoten, 2017. (in Japanese).
3. Minamata City Council. The Declaration of the Development of City that Treasure Environment, Health, and Social Welfare, June 1992.
4. Minamata City. Declaration of Creating an Environmental Model City. November 1992.
5. Minamata City. Minamata City Environmental Report, 2017. March 2019. (in Japanese). https://www.city.minamata.lg.jp/kankyo/kiji003505/3_505_up_oczhxj1p.pdf.
6. Global Environment Centre Foundation. Eco-towns in Japan. UNEP. June 2005. https://wedocs.unep.org/bitstream/handle/20.500.11822/8481/Eco_Towns_in_Japan.pdf?sequence=3&%3BisAllowed=
7. de Leeuw E, Simos J. Healthy cities: the theory, policy, and practice of value-based urban planning. New York: Springer; 2017.
8. Environmental Health and Safety Division, Environmental Health Department, Ministry of the Environment, Japan. Lessons from Minamata Diseases and Mercury Management in Japan. Tokyo: Ministry of the Environment Japan; 2013. https://www.env.go.jp/chemi/tmms/pr-m/mat01/en_full.pdf.
9. Minamata City. Minamata Declaration for Creating a Zero Waste City. 2009.

Chapter 17
Divorcing from Plastics for a Sustainable Future Society

Tamie Nakajima

Abstract Plastic products, such as containers and packaging, have deeply penetrated our daily lives, and their production has been increasing year by year. Plastic marine pollution is also spreading in tandem with plastic products and is now a global environmental problem. Plastics in the ocean are fragmented by ultraviolet light and physical force and are the origin of microplastics. Plastics also contain toxic substances, such as plasticizers, that have reproductive and developmental toxicity. The importance of risk assessment is paramount regarding the effects of microplastics containing these toxic substances on the ecosystem and human health. It is necessary to reduce the number of single-use plastics first to transform from the plastic society we have become into a sustainable society and achieve SDGs by 2030.

Keywords Containers and packaging · Microplastics · One-way plastics · Plastics Plasticizers · Polychlorinated biphenyls · Risk assessment · Sustainable development goals (SDGs)

17.1 Introduction

The word "plastic" is an adjective derived from the Greek word that means "to trace" [1]. The invention of celluloid in 1869 marked the beginning of plastics. In the twentieth century, the term "plastic" has been generally used for new synthetic materials made of polymers based on carbon and hydrogen. In 1907, Bakelite, a

---

T. Nakajima (✉)
Nagoya University, Nagoya, Japan

College of Life and Health Sciences, Chubu University, Kasugai, Aichi, Japan
e-mail: tnasu23@med.nagoya-u.ac.jp

T. Nakajima et al. (eds.), *Overcoming Environmental Risks to Achieve Sustainable Development Goals*, Current Topics in Environmental Health and Preventive Medicine, https://doi.org/10.1007/978-981-16-6249-2_17

hard plastic used in radios, telephones, and household appliances, was created. After the booming of petroleum complexes in 1960, plastics such as polyethylene began to be produced, and plasticizers, curing agents, colorants, aromatics, and stabilizers followed, resulting in the development of a great variety of plastics.

Plastics are used in many fields because of their light weight and convenience. Various types of plastics are used for food packaging. For example, polyethylene terephthalate is used in plastic bottles. Polyethylene is used for freezer bags, polypropylene, for food packaging, and polyvinyl chloride (PVC), for medical tubes and bags. Polystyrene is used in fish and meat trays and disposable food containers, and containers for cup noodles. Epoxy resin is used for coating the inside of cans; acrylic resin, for tableware and toys such as blocks; polytetrafluoroethylene, for fluoropolymers and frying pans; melamine resin, for melamine dishes, forks, and spoons, cooking sheets, and container lids. Silicone is used in kitchen utensils. Other materials include polyurethane, which is used in household items, such as mattresses, hoses, sportswear, and tights, and polycarbonate, used in CDs, DVDs, and eyeglasses. Notably, polyethylene and polypropylene are also used in nonwoven masks, which became a necessity item during the COVID-19 pandemic.

The term "plastic society" has not suddenly come into use. In the 1990s, plastics, especially chlorinated ones, attracted attention as a source of dioxin in Japan. Because of its convenience, priority was placed on plastic production, but no attention was paid to the disposal method. At the time, it was burned for disposal, which produced dioxin during the process. This became a major social issue, leading to a boycott of vegetables and other agricultural products near incineration plants. As a result, incinerator management was introduced by the government, and dioxin risk management has been issued. However, over time, marine pollution caused by illegally discarded plastics has become a serious problem, and it has progressed into microplastic pollution.

In this paper, first, plastics, their types, and uses are outlined. Then, marine pollution by plastics and microplastics is examined. Third, the hazards and risk assessment of plasticizers in plastics are stated. Fourth, the issue of plastics pertaining to the action plans of the 2030 agenda for sustainable development is evaluated, and finally, we state the need to reduce the weight of one-way plastic to achieve six goals of the SDGs.

17.2 Applications and Production of Plastics in Japan

Regarding the applications of plastics in Japan in 2019, containers and packaging film sheets (containers 14.8% and packaging film sheets 20.6%) accounted for one-third of total plastics (Fig. 17.1). Other applications were film sheets other than for packaging (18.5%), machinery, equipment, and parts (12.1%), pipes and fittings (7.6%), daily necessities and miscellaneous goods (5.0%), building materials (4.7%), foam products (4.2%), and others. Comparing the number of plastic containers and packaging disposed of per capita in each country, Japan ranks second after the United States.

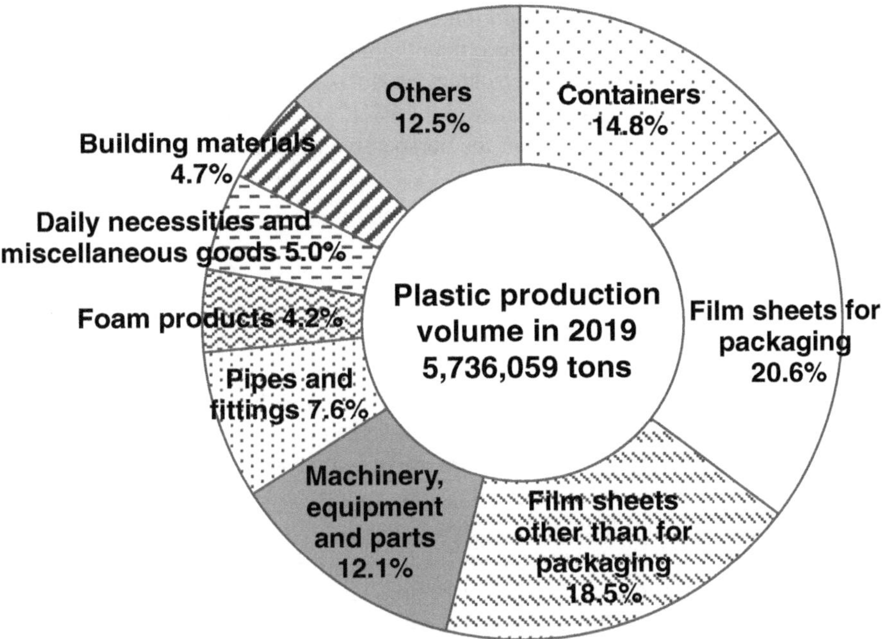

Fig. 17.1 Plastic applications. Source; Vinyl Environmental Council

Globally, more than 8,300,000,000 tons of plastic have been produced since 1950, and 6,300,000,000 tons of that have been disposed of as waste [2]. Seventy-nine percent of the collected plastic waste is either landfilled or abandoned in the ocean. Only 9% of plastic is recycled. At the current rate, it is estimated that more than 12,000,000,000 tons of plastic will be landfilled or thrown away into the environment by 2050.

Changes in the production of plastics in Japan over the past 20 years show that the production of containers has doubled [3]. However, the total production volume has not substantially changed because reinforced products are no longer included, and the production of foam artillery, construction materials, daily necessities, machinery and equipment parts, and plates has decreased. Since the production volume of containers is expected to increase linearly in the future, limiting the production of these products will certainly be the key to controlling the total amount of plastic emissions.

17.3 Advantages and Disadvantages of Plastics

Plastic extends the shelf life of food products and, because of its light weight, it also facilitates their transportation. This advantage has led to its widespread use as containers and packaging products [2]. Plastic containers have doubled even in the past 20 years, and it is projected to double in the next 20 years. In other words, its increase is almost linear. If this situation continues, it is predicted that the amount of plastic in oceans will increase more than the number of fish by 2050. Moreover, by 2050, the share of plastics in petroleum products will increase by 20%, and the share of plastics in the carbon balance, by 15% [2]. The production of plastics is related to global warming, and from the perspective of climate change, plastic pollution in the ocean is a serious problem.

17.4 Marine Pollution by Plastics

First, it is assumed that the illegal dumping of plastic, like littering, will lead to ocean pollution. The total production of plastics in Japan has been flat for the past 20 years [3]; however, the COVID-19 pandemic has increased the demand for take-out plastic containers, protective masks (made of polypropylene and polyester), and medical protective equipment (made of polypropylene and polyethylene). The total production of plastic products may increase due to the increased demand for containers for these demands. Further, the production of plastic products in other countries is also expected to increase in the future. The amount of plastic transported from Southeast Asia and China by the Kuroshio Current is expected to increase. In 1973, plastic was detected in the digestive tracts of seabirds in Canada [4]. Although there have been sporadic reports of similar findings since then, they have not been acknowledged until the twenty-first century. Increased pollution and its effects on seabirds and sea turtles led to the use of the term "plastic ocean." In other words, plastic marine pollution has become a matter of life and death for marine life. In addition to the negative impacts on the landscape and coastal habitat, it may also hurt tourism and fishing.

17.5 Environmental Pollution by Microplastics

Microplastics are plastics smaller than 5 mm in length and are classified into primary and secondary microplastics. The former are microbeads in cosmetics and face washes, resin pellets (small, bead-like plastics used as raw materials in the manufacture of plastic products), and fertilizer-coated slow-release fertilizer capsules. The latter are fragments of plastic and scraps of clothing fibers. Secondary microplastic pollution from plastic debris is common in the seas around Japan. Microplastics in oceans adsorb toxic chemicals, such as PCBs, and act as their carriers [5], explained in Chap. 10. Therefore, even if the concentrations of PCBs and other toxic chemicals are low in the seas around Japan, microplastics may adsorb them in high seas and carry them to the seas around Japan by the Kuroshio Current.

Many toxic chemicals are used in plastics, such as plasticizers and hardeners. When they are micronized, they are more easily ingested by fish, and the toxic chemicals are more easily absorbed by the organism. If the toxic chemicals are bioaccumulated, they may have an impact on humans. Although many researchers have clarified the toxicity of these toxic chemicals alone, little research has been done on the ecological and biological effects of microplastics with toxic chemicals.

Recently, Isobe et al. measured the number of microplastics in the Pacific Ocean [6]. They showed the distribution of microplastic concentrations in the Pacific Ocean, the lowest being 10 mg/m^3 and the highest, 1000 mg/m^3. Seasonal variations occurred in the concentration of microplastics in the Pacific Ocean, with low concentrations in February and high concentrations in August, and high-concentration areas (hot spots) appeared in the seas around Japan in August. This suggests the need for early action to reduce marine microplastics.

Interestingly, microplastics have been detected in the urban air and even in indoor air. Wright et al. detected 575–1008 (m^2/d) microplastics in Central London in February 2018 [7]. The shape of the microplastics was mainly fibrous microplastics accounting for 92% of the total. Moreover, in Japan, the digital edition of the Asahi Shimbun reported the detection of polyethylene and polypropylene of tens to hundreds of micrometers [8]. Moreover, polyethylene was also detected in the ice of trees in the mountains of Kyushu. These results show that microplastic contamination is a global problem that can be detected not only in the ocean but also in the atmosphere.

In the case of microplastics, in addition to the effects of the material, the effects of toxic substances contained in the material must also be considered. For example, styrene trimer, a byproduct of polystyrene resin, disrupts thyroid hormones [9]. In addition, toxic substances that are adsorbed by microplastics also cause adverse effects. For example, perfluorooctanoic acid (PFOA) has been shown to cause liver damage in mice [10, 11] and has adverse effects on the lipid homeostasis of women with low-birth-weight infants [12]. As described above, the effects of toxic chemicals have been extensively studied, but there has been little research on the forms contained in or adsorbed toxic chemicals on microplastics.

17.6 Health Hazards of Hazardous Chemicals Included in Plastics

17.6.1 Effect of Styrene Trimmer on Thyroid Hormones

Styrene trimmer is a byproduct of polystyrene resin manufacturing. In the past, polystyrene resin was used to make cup noodle containers, and concerns remained related to the leaching of styrene trimers, monomers, and dimers from these containers. It has been reported that styrene trimers administered to mice increased thyroid hormones [9]. Recently, plastic containers for cup noodles have been replaced with paper containers. Our research on the toxicity of various chemicals may be a nuisance to companies, but it has led to the development of new containers. It seems that toxicity research is having a positive impact on pollution-related problems.

17.6.2 Effect of Plasticizer Diethylhexyl Phthalate on the Next Generation

Plastic is inherently hard, so tubes and blood bags used in hospitals are made of softened plastic using plasticizers, such as di-2-ethylhexyl phthalate (DEHP) (Fig. 17.2), which is no longer used in medical tubing, but about 20 years ago, it was used in 40% of medical materials and single-used groves. Since plasticizers are fat-soluble, they are easily eluted when products are disinfected with alcohol or come into contact with oil. In 1999, the concentration of DEHP in convenience store lunch boxes and retort pouch foods was found to exceed the tolerable daily intake (TDI, that is the daily intake of a substance estimated to have no adverse effect on human health when consumed continuously over a lifetime) because of PVC gloves used during the cooking process [13]. On June 14, 2000, the Ministry of Health and Welfare issued a notice to prevent their use, and the DEHP concentration in convenience store lunch boxes was subsequently reduced to a much lower level. DEHP has a structural formula consisting of two ethylhexyl alcohol esters bonded to a phthalic acid skeleton and is fat-soluble. According to recent data, DEHP accounts for 42% of plasticizers. The use of DEHP is gradually decreasing as the toxicity evaluation of DEHP progresses. On the contrary, the use of diisononyl phthalate (DINP) has been increasing because DINP is less toxic than DEHP. In addition, phosphoric acid, adipic acid, and epoxy plasticizers are also being used.

DEHP-containing tubes were used in the hospital around the 2000s; therefore, we investigated the concentration of DEHP exposure in patients [14]. We compared the concentration of mono-ethylhexyl phthalate (MEHP), a major active metabolite of DEHP, in the serum of 28-day-old infants who had been tube-fed since soon after birth because of their medical condition with those of 28-day-old breastfed infants and dialysis patients. Surprisingly, the concentration in the tube-fed patients was five times higher than that in the breastfed. On the other hand, since a lot of tubing was used during dialysis, we expected that the serum MEHP concentration of these patients would be the highest but found that it was about the same as that of breast-fed children. The fact that the DEHP concentration was much higher in tube-fed

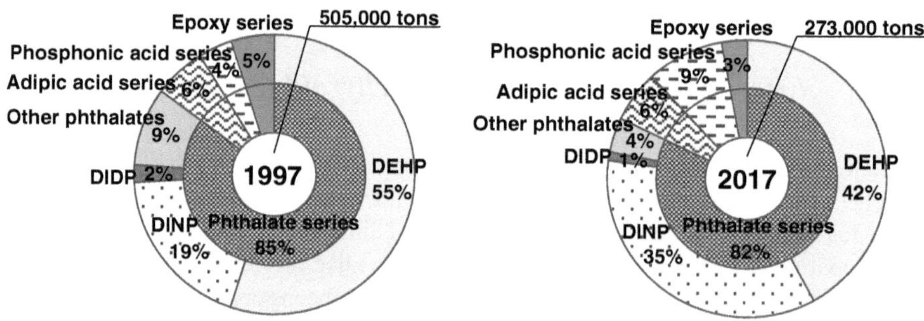

Fig. 17.2 Typical plastic plasticizers. *DEHP* di-2-ethylhexyl phthalate, *DINP* diisononyl phthalate, *DIDP* diisodecyl phthalate. Source; Vinyl Environmental Council

children than that in dialysis patients attracted much attention. On October 17, 2002, the Ministry of Health, Labor, and Welfare issued a warning to the Japan Medical Association, municipalities, and Pharmacists' Associations that DEHP, a plasticizer eluted from PVC medical devices, could exceed the TDI at that time. About 8 years later, we measured DEHP elution from medical tubing and vinyl-based bags but could not detect DEHP, confirming that it was not being used at all.

The effects of DEHP on the next generation have been studied. Wild-type, peroxisome proliferator-activated receptor (PPARα)-deficient, and human PPARα pregnant mice were fed with diets containing 0%, 0.01%, 0.05%, and 0.1% DEHP [15]. This method of exposure was converted to daily DEHP intake: 0, 10 ~ 12, 55 ~ 64, and 110 ~ 145 mg/kg/day. We then examined the numbers of fetuses and births per litter and the numbers of survivors on gestation day 18 and birthday 2, respectively. In PPARα-deficient mice, there was no effect of DEHP administration on these parameters. In the fetuses of human PPARα mice, there was no effect of DEHP administration on the fetuses, but the number of pups at day 2 and the number of surviving pups were reduced at a dose of 0.1%. In contrast, no toxic effects were observed in the fetuses of wild-type mice (murine PPARα) at any dose, but the number of surviving pups was reduced 2 days after birth in the 0.05% and 0.1% dose groups. These results indicate that the effects of DEHP on the fetus and newborn pups are PPARα-dependent, and since the activity of PPARα is mouse > human, the toxicity is stronger in wild-type mice. Since human PPARα is expressed only in the liver, the role of PPARα in the liver plays an important role in pup survival, suggesting that lipid homeostasis in the mother's liver has a significant impact on DEHP-induced pup survival.

Next, we will discuss one way to assess the health risk from these results (Fig. 17.3): based on the measured intake of DEHP-containing food, the DEHP dose that had no adverse effect was 11 mg/kg/day. Therefore, the no observed adverse effect level (NOAEL) was 11 mg/kg/day. Dividing this by 100 (uncertainty factor, UF) we obtained a TDI of 0.11 mg/kg/day. The divisor 100 is called the uncertainty factor and is used to set a safe TDI to account for the uncertainty in the NOAEL. Figure 17.3 shows the dose on the horizontal axis and the response on the vertical axis.

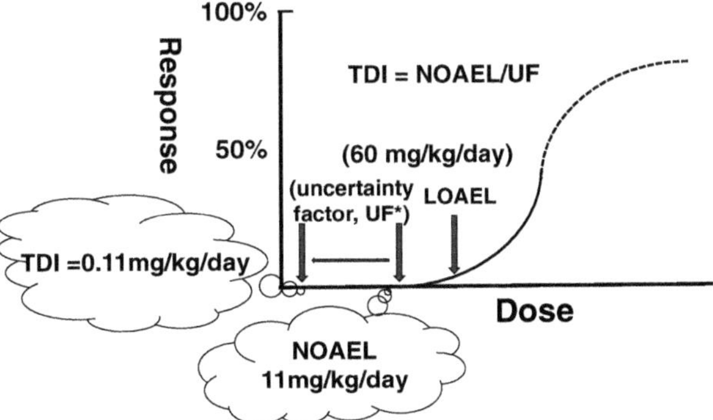

Fig. 17.3 Dose–response relationship and risk assessment of DEHP. *DEHP* di-2-ethylhexyl phthalate; *NOAEL* no observed adverse effect level, *LOAEL* lowest observed adverse effect level, *TDI* tolerable daily intake. *Species and individual differences 10 × 10

However, researchers in the Netherlands reported a shortening of the anogenital distance and a decrease in the weight of reproductive organs at birth in male rats due to DEHP [16]. The TDI of DEHP was calculated as 0.03 mg/kg/day. Since the TDI of DINP was 0.15 mg/kg, it was less toxic than DEHP. According to the latest report, the amount of DEHP ingested in Japan is about 5 µg/kg/day in 2004 [17], which is one-fifth of DEHP TDI, although the data is a little old. As this is a toxicity study of DEHP, a plasticizer contained in plastics, studying the combined effects of plastics and plasticizers or plastics and additives is also required.

17.7 SDGs: An Invitation to a Future Society

SDGs that are closely related to plastic water pollution include the following:

1. Target 9 of Goal 3, "health and well-being for all": By 2030, the number of deaths and illnesses caused by hazardous chemicals and air, water, and soil pollution will be substantially reduced.
2. Target 3 of Goal 6, "safe water and toilets for all": By 2030, improve water quality by reducing pollution, eliminate the release of harmful chemicals, reduce the proportion of untreated wastewater by half, and significantly increase recycling and safe reuse on a global scale.

   Target 6 of Goal 11, "create livable cities": By 2030, reduce adverse environmental impacts per capita in cities, including those resulting from special attention to air quality and general and other waste management.

   These three goals all involve significantly reducing the release of hazardous chemicals, plastics, and other wastes into the air and water, thereby leaving a high-quality environment for the future. Efforts to reduce microplastics, which are carriers of toxic chemicals, are equally necessary.
3. Target 4 of Goal 12, "responsibility to create, responsibility to use": By 2020, in accordance with agreed international frameworks, achieve environmentally sound management of chemicals and all wastes throughout the product life cycle, and significantly reduce the release of chemicals and wastes into the air, water, and soil to minimize their adverse effects on human health and the environment.

   Researchers in the field of industrial hygiene need to consider not only the health of workers but also the management of materials produced. In addition, the disposal of plastic into the ocean should also be reconsidered and reduced.
4. Goal 13, "specific policies on climate change"

   This seems to be the highest priority globally. Witnessing the disasters in eastern Japan caused by Typhoons 15 and 19 that hit the country in 2019, we are keenly aware of the urgent need for concrete policies. Of course, as mentioned earlier, the involvement of plastic production is also significantly related. In 2020, Japan's prime minister announced in his policy speech that Japan would reduce greenhouse gas emissions to practically zero by 2050, with a "virtuous cycle between the economy and the environment" as a pillar of the growth strategy.

5. Target 1 of Goal 14, "protect the richness of the oceans": By 2025, prevent and substantially reduce all types of marine pollution, including marine litter and eutrophication, especially pollution from land-based activities.

17.8 Taking Action for Plastic Reduction

Figure 17.4 shows the plastic material flow chart in Japan (2018). In our country, material recycling accounts for 23%, and chemical and thermal recycling accounts for 61% of the total recycling. However, chemical and thermal recycling are not truly recycling, and they promote global warming. Looking at material recycling in 2013, most of it (83%) was exported to Asian countries, mainly China, and domestic recycling was only about 16%. The export to China was banned in 2017, and material recycling is now at a standstill [2]. In terms of national action, the Osaka Blue Ocean Vision calls for reducing new pollution from plastic waste in the ocean to zero by 2050. In response, the Science Council of Japan has advised that "the urgent task is to develop and implement technologies to collect and treat existing plastics as soon as possible" [3]. The first domestic milestone is to reduce the amount of one-way plastic, and the second is to reuse and recycle it. Since plastic resources will gradually decrease, the reduction plan is a must. In the case of reuse, it is necessary to consider that harmful substances do not leach out, even after multiple uses. Thus, the next step is the use of recycled and biomass plastics.

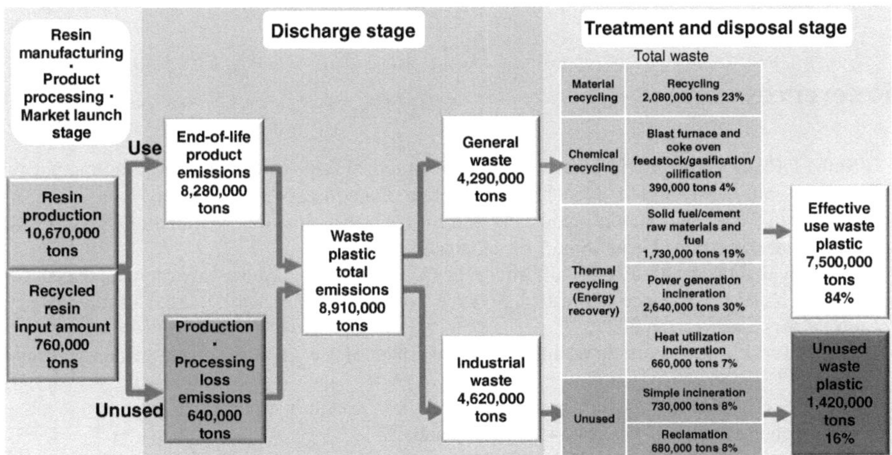

Fig. 17.4 Material flow of plastics (2018). Source; Ministry of the Environment website https://www.env.go.jp/council/03recycle/20201120t2.pdf

The Tokyo Metropolitan Government was looking for universities to campaign for the reduction of one-way (disposable) plastics in June 2019. The purpose of calling on universities in this campaign is to raise environmental awareness among students, reduce the amount of waste produced by the university, and contribute to their environmental protection activities. As an example of action at a university, I encouraged the students to think about what actions should Chubu University take. I showed the video "Plastic Ocean" to the students and asked them what they thought they should do after watching the video. The students stated "I will bring a water bottle tomorrow instead of using a plastic bottle" and "I will refuse to use a bag when I go to the checkout counter." There was also a suggestion that we should reduce the use of disposable plastics. I believe education is the most important factor for achieving SDGs. Chubu University has a very long history of education for sustainable development, so I encourage continuous efforts along these lines and also focusing on building science literacy.

17.9 Conclusion

As René Dubos stated in his book *Courtship of the Earth* about 40 years ago, we must "think globally, act locally." One of my students also wrote, "it is important for each and every one of us to take action to curb plastic pollution of the ocean." The 17th goal of SDGs is partnership, and it is important for each of us to collaborate with the rest of the world to multiply our one-way plastic reduction efforts. (This chapter was originally presented at Chubu University on December 11, 2019).

Acknowledgments I would like to thank Mrs. Chizuru Narishima for her support in creating the figures.

References

1. Sinha J, Plamondon C. Life without plastic: the practical step-by-step guide to avoiding plastic to keep your family and the planet healthy. Salem: Page Street Publishing; 2017.
2. Ministry of the Environment. Domestic and international situation surrounding plastics (3rd document collection) Plastic Smart (in Japanese).
3. Ecological and Health Effects of Microplastic Pollution of Water Environment Need for Research and Governance of Plastics. Science Council of Japan April 7, 2020 http://www.scj.go.jp/. Accessed 15 Jan 2021.
4. Rothstein SI. Plastic particle pollution of the surface of the Atlantic Ocean: evidence from a seabird. Condor. 1973;75:5.
5. Takada H. Marine plastic pollution and its countermeasures trends in science. Gakujyutsunodoukou. 2019;24:44–9. (in Japanese)
6. Isobe A, Iwasaki S, Uchida K, et al. Abundance of non-conservative microplastics in the upper ocean from 1957 to 2066. Nat Commun. 2019;10:417. https://doi.org/10.1038/s41467-019-08316-9.

7. Wright SL, Ulke J, Font A, et al. Atmospheric microplastic deposition in an urban environment and an evaluation of transport. Environ Int. 2020;136:105411.
8. Microplastics also found in morning air in Fukuoka City, Asahi Shimbun Digital November 19, 2019 (in Japanese).
9. Yanagiba Y, Ito Y, Yamanoshita O, et al. Styrene trimer may increase thyroid hormone-levels via down-regulation of the aryl hydrocarbon receptor (AhR) target gene UDP-glucuronosyltransferase. Environ Health Perspect. 2008;116:740–5. https://doi.org/10.1289/ehp.10724.
10. Nakamura T, Ito Y, Yanagiba Y, et al. Microgram-order ammonium perfluorooctanoate may activate mouse peroxisome proliferator-activated receptor alpha, but not human PPAR alpha. Toxicology. 2009;265:27–33. https://doi.org/10.1016/j.tox.2009.09.004. Epub 2009 Sep 12
11. Nakagawa T, Ramdhan DH, Tanaka N, et al. Modulation of ammonium perfluorooctanoate-induced hepatic damage by genetically different PPARα in mice. Sotoumi Arch Toxicol. 2012;86:63–74.
12. Kishi R, Nakajima T, Goudarzi H, et al. The association of prenatal exposure to perfluorinated chemicals with maternal essential and long-chain polyunsaturated fatty acids during pregnancy and the birth weight of their offspring: The Hokkaido study. Environ Health Perspect. 2015;123:1038–45. https://doi.org/10.1289/ehp.1408834.
13. Tonogai Y. A study on the actual situation of food contamination and intake of phthalate esters and phenols. http://www.nihs.go.jp *'edc' houkoku12 '12- tonogai' tonogai-soukats.*
14. Ito Y, Kamijima M, Nakajima T. Di(2-ethylhexyl) phthalate-induced toxicity and peroxisome proliferator-activated receptor alpha: a review. Environ Health Prev Med. 2019;24:47. https://doi.org/10.1186/s12199-019-0802-z.
15. Hayashi Y, Ito Y, Yamagishi N, et al. Hepatic peroxisome proliferator-activated receptor α may have an important role in the toxic effects of di(2-ethylhexyl) phthalate on offspring of mice. Toxicology. 2011;289:1–10. https://doi.org/10.1016/j.tox.2011.02.007.
16. Christiansen S, Boberg J, Axelstad M, et al. Low-dose perinatal exposure to di(2-ethylhexyl) phthalate induces anti-androgenic effects in male rats. Reprod Toxicol. 2010;30:313–21. https://doi.org/10.1016/j.reprotox.2010.04.005.
17. Itoh H, Yoshida K, Masunaga S. Evaluation of the effect of governmental control of human exposure to two phthalates in Japan using a urinary biomarker approach. Int J Hyg Environ Health. 2005;208(4):237–45. https://doi.org/10.1016/j.ijheh.2005.02.00.

Chapter 18
Issues of Environmental Risks in the Anthropocene

Chiho Watanabe

Abstract During the last seven decades, characterized by the *Great Acceleration*, we have experienced two substantial changes associated with the area of environmental risks. First, we are forced to be aware of the "size" of the globe, which is symbolized by the proposal of *Planetary Boundaries* (PBs) concept. Second, the number and amount of chemicals produced, released into our environment continue to grow throughout these decades. The concept of PBs urges us to rethink about environmental risks in the context of the *finite* world. Also, we need to know PBs are primarily devised for earth systems (like climate system) and are necessary but not sufficient conditions for the survival and well-being of humans (society). Growing number of chemicals had led to the idea of *exposome*, aiming at grasping whole exposures to chemicals and other environmental factors. To solve the issues of combined exposures, however, a systematic/theoretical approach like *Adverse Outcome Pathway* needs to be developed. These scientifically intriguing and practically important challenges will extend the scope of environmental risks to both directions: global and molecular.

Keywords Great Acceleration · Anthropocene · Planetary boundaries · Planetary health · Exposome · Adverse outcome pathway (AOP) · Combined exposure

---

C. Watanabe (✉)
School of Tropical Medicine and Global Health, Nagasaki University, Nagasaki, Japan
e-mail: chiho.watanabe@nagasaki-u.ac.jp

T. Nakajima et al. (eds.), *Overcoming Environmental Risks to Achieve Sustainable Development Goals*, Current Topics in Environmental Health and Preventive Medicine, https://doi.org/10.1007/978-981-16-6249-2_18

18.1 Two Major Changes during the Last Seven Decades

Most chapters of this book focus on the events occurred during the most recent seven decades, i.e., from the 1950s to 2010s, when the world as well as Japan experienced a tremendous change of environment, society, and economy. Roughly summarizing the situation in Japan, many issues related to environmental pollution erupted or surfaced in the 1950s (although the existence of some diseases like Itai-itai disease had been known well before this period), the 1970s was an era of rapid economic growth and the peak of environmental pollution issues, then new types of environmental contamination like dioxins and endocrine-disrupting chemicals emerged in late 1980s, followed by "global" issues starting from the global warming, ozone depletion, etc. Thus, environmental issues that Japan (and most probably, rest of the world also) was confronted with have been continuously changing. There are two major changes relevant to the scope of this book, the environmental risks.

First, currently, 7.8 billion people are living on the globe, while only 2.5 billion were in 1950, tripled in the last 70 years, which chapters of this book mostly cover. The last half of the twentieth century is an era of the so-called *Great Acceleration*, in which human society rapidly increased the sizes of many activities, including carbon emission, fertilizer use, water use, urbanization, and population, etc. [1]. Anthropocene [2], the era of human activities destroying and perturbing the stability of earth, is the result of these ever-increasing activities. It follows that the relationship between human society and the globe has been changed during these decades; we used to have a planet that was large enough, but we no more have it. If the size of the planet were far bigger, probably we do not have to worry about the Great Acceleration. Great Acceleration, Anthropocene, and planetary boundaries (described later) all come from the fact that we are living in a *finite* world, and more importantly, we begin to feel the *size* of the globe.

Second, during these decades, number and quantity of chemicals/materials drastically increased; Material extraction in 1950 was around 7 billion tons (Bt), which grew rapidly, reaching 26.7 Bt in 1970, and 92.1 Bt in 2017 [3]. Production of organic chemicals increased from seven million tons (Mt) in 1950 to 250 Mt. in 1985 [4]. Number of entries in the Chemical Abstract Service (CAS) Registry started in 1965 was 3 M in 1975, 10 M in 1990, and reached 100 M in 2015. Unfortunately, because of this explosion of chemical diversity, fragmentation of research/regulatory communities has developed in the area of environmental risk of chemicals. Recently, a multinational group of researchers published a paper advocating the establishment of a new international science-policy initiative, a chemicals version of IPCC, so to speak, to overcome this fragmentation [5].

Before discussing future environmental risks, definition of the term, *environmental risk*, may be worth discussing briefly. The term is often considered as risks associated with chemical substances (at least in Japan). USEPA defines *their* risk as risks to humans and ecological receptors from chemical contaminants and *other stressors* [6]. EC includes risks not directly associated with chemical pollution; for example, their 2013 report covers such issues like climate change, flood, GM crops, invasive alien species, mobile phones, and nanotechnology [7]. Hence, it would be safe to extend the definition of the environmental risks as including both chemical and non-chemical factors.

18.2 Recognizing the Size of our Finite World

18.2.1 Planetary Boundaries

The recognition of living in a finite world is represented by the notion of the *Planetary Boundaries* (PBs), which warns us that including *climate change*, there are nine areas that require careful monitoring (namely, *climate change, biosphere integrity, land-system change, freshwater use, biochemical flows, ocean acidification, atmospheric aerosol loading, stratospheric ozone depletion, novel entities*) and, at least for some of them, urgent actions to achieve sustainability of the earth [8, 9]. Each of these nine areas is thought to have a respective planetary boundary, which once exceeded fundamental changes in the earth subsystem will result. In the second version of PBs [9], two *core* PBs are identified, i.e., *climate change* and *biodiversity*, which requires us to keep either the atmospheric concentration of CO_2 below 350 ppm or the number of vanishing species less than 10 per one million species; substantial degradation of biodiversity will weaken the resilience of earth subsystem (through reduced forest area resulting in reduced CO_2 absorption and flood-prone land, for example).

18.2.2 Risks of Chemicals in the Context of the Planetary Boundaries: Local-Global Connection

Risks imposed by chemical substances is one of such PB areas, which is referred to as "*novel entities*" in the second (latest) version of PBs paper, since many artificial substances are "new" to the earth/nature and may be hard to deal with for natural ecosystems [9]. Together with the other two areas (i.e., *aerosol* and *"functional" biodiversity*), there is no known boundary for this area; neither the quantitative level of boundary nor the feasibility of defining the boundary is known. This is simply due to the huge number of chemicals and wide range of their nature, although some groups suggested that such a boundary might have been already transgressed [10].

Apart from *novel entities*, some of the other PBs areas are related to the amount and localization of chemical substances. That is, *climate change* (excess CO_2 in the atmosphere), *biogeochemical flows of nitrogen and phosphorus*, *atmospheric aerosol* (PM2.5, etc.), *ozone depletion* (CFCs in stratosphere) are related to the localization and/or flow of respective elements. In addition, two other areas, *freshwater use* (related to contamination) and *biosphere integrity*, are affected by chemical pollution. Therefore, controlling the production, use, and release of chemicals in a broad sense is associated with at least seven out of nine PB-areas, constituting a critical point to achieve the sustainability of the globe.

Since we become aware of the finite nature of the world, conventional risk evaluation can be reconsidered in the context of PBs. Conventional indicators like no-observed-adverse-effect levels (NOAELs) or Benchmark Doses (BMDs) have little to do with the size/volume of "space," where the chemical(s) of concern is present (except for the size/volume of individual biological system -like individual/organ/cell, etc.). It might be worth considering how these *local* (on site/in situ) indicators are related with PBs, which is global. There are some attempts, which try to show the connection between "local" with global. For example, marine plastic pollution has been evaluated if it meets the criteria for being a threat to PBs by affecting *climate change* and *biodiversity*, the two "core" PBs [11]; also, by estimating the tolerable load of various environmental stressors (like climate change, land use, eutrophication, etc.), quantitative relationships between various environmental thresholds (like PBs) and corresponding human activities generating such stressors (CO_2 emission, for example) were examined [12]. Elucidating environmental behavior of elements/substances may help to quantify such connections: for example, local release of phosphorus may cause eutrophication but support food production, which bears more global effects [13]. Climate change may affect the environmental behavior of mercury at global as well as local scales [14].

18.2.3 Planetary Boundaries May Not Be Sufficient for Human Health and Well-being

As mentioned above, PBs primarily concern functions of the earth sub-systems. Keeping our activity within PBs is a necessary condition for human health but may not be a sufficient condition. Various types of global environmental changes ranging from climate change to biodiversity – including chemical pollution – affected and will affect the occurrence and prevalence of many non-communicable diseases through various pathways [15]. Mental health has been identified among the significant consequences of climate change [16]. Climate change within +2 °C can be linked to increased production of Aflatoxin B, one of the most toxic mycotoxins in maize in Europe [17]. Global warming has already led to an increased chance of harmful algal blooms [18]. On the other hand, based on analyses across 37 countries, excessive environmental loading in terms of CO_2, phosphorus, and land use associated with national recommendations for diet (NDR) has been identified in some (poorer middle-income) nations; optimizing NDR could alleviate the environmental impact [19]. These examples show possible links among the physico-chemical environment, biosphere, and human health/well-being; quantitative investigations are needed to clarify the relationship between PBs and health/well-being.

18.3 Too Many Chemicals, Too Many Interactions

Conventional environmental risk assessment mostly focuses on a single agent or condition as the causative factor of certain adverse health outcome. Growing number of chemicals, as described at the beginning of this chapter, made it impractical to examine such agent–outcome relationship exhaustively. In addition, since the effects of vast number of chemicals often converged into limited number of biological systems, it may become gradually less important to elucidate such one-to-one relationships. Instead, estimating/predicting the total effect of group/mixture of chemicals became an important but difficult challenge for chemical risk assessment. Risks of other non-chemical environmental factors would be imposed on/interact with risks of chemicals, which further complicates the situation. Two potentially effective approaches to tackle problems of this kind, *exposome* and *adverse outcome pathway* (AOP), will be described below.

18.3.1 Combined Exposure among Chemicals and with Non-chemical Factors: (Including Exposome *Approach)*

Exposome has been first proposed by Wild [20] as a concept in environmental epidemiology. The term *exposome* has been created to express exposure equivalent of *genome*: as genome is encompassing entire genetic material, the exposome is encompassing entire exposures. Thus, in the study employing exposome approach, *any* type of exposure to environmental factors (not limited to chemical exposure) would be recorded throughout the life of a person. Then, the cumulative record of the exposures will be analyzed in relation to many health outcomes of the person. While this approach sounds too ambitious, and it is definitely time-, money-, labor-intensive, a pilot multi-national study, HELIX study, consisted of six cohorts with over 30,000 participants was conducted in Europe in 2014–2019 [21], which is now extended in scale with more elaborated methods (ATHLETE study, 2019). Some initial results of the HELIX study have been published, where clusters of environmental factors sharing similar exposure pattern were identified [22], while issues of statistical methods were also identified [23].

A couple of large cohort studies can be found in various parts of the world: JSEC (Japanese Study of Environment and Children) is among the largest, involving around 100,000 mother–baby pairs, which is planned to do follow-ups until the participants will reach the age of 13. This study already generated interesting results taking advantage of the size of the cohort, with a conventional statistical approach (e.g., [24]).

Sometimes health outcome of an exposure to chemicals differs by sex, age, and/or genetic make (i.e., genetic susceptibility) of the individual/population. For example, skin manifestations of inorganic arsenic were far more severe among males than females [25], prenatal exposure to methylmercury results in severe neurological consequences, which is substantially different from those found after adult exposure. In addition to these biological attributes, many external (environmental) factors would affect or interact with the risk of chemicals: e.g., global warming, over/under-nutrition, emerging infectious diseases.

Current risk evaluation has already incorporated genetic factors and biological attributes but paid less attention to the temporal and spatial variation of the environmental conditions. To make risk evaluation more useful across nations/populations and under rapidly changing environmental condition, the substance needs to be evaluated under various environmental settings, or interaction of the adverse effect of concerned chemical with other environmental factors should be examined. This can be regarded as an extended version of the chemical-chemical interactions as noted above.

18.3.2 Systematic/Mechanistic Understanding of Adverse Outcomes

Given the number of chemicals and non-chemical environmental factors, it is too obvious that the number of the combinations would be huge, and exhaustive assessment of combinations is impossible. A practical solution for this problem is to pick up some typical/representative combinations based on collected facts or a priori assumptions and test those combinations. For example, a group of chemicals sharing similar chemical structures (or similar usage) could be treated as a whole to collapse the number of the (combination of) chemicals that need to be tested. Another approach, *adverse outcome pathway* (AOP), is offering a more systematic solution, trying to elucidate the pathway starting from the chemical's entry to the exit, in which the pathway is expressed as the series of modules. Combination of exposome and AOP would provide a very powerful tool [26]. While this combined approach appears as time-money-labor intensive, using existing knowledge such as QSR and/or utilizing AI would make the construction of pathway map feasible [27, 28].

18.4 Future Directions

As discussed so far, the two important changes observed in the last seven decades, the period covered by this book, may introduce new tasks in the area of environmental risks. The first one, recognition of the finite world, particularly identification of PBs, indicates the need for connecting current *local* (or, on site – in situ) risk scheme with risks associated with *global level*. The second one, increasing the number and amount of chemicals and recognition of many internal (biological) and external interacting factors altogether suggest the need for systematic *and* molecular-level understanding of adverse biological effects, which has not been extensively explored so far. Thus, the scope of environment risk study needs to be extended to two opposite directions, *global* and *molecular*. It also needs to be noted that PBs may not be a sufficient condition for human health and well-being [12]. Therefore, in addition to PBs, we need *planetary health*, which emphasizes the importance of earth and biosphere as supporting human health and civilization, as a core concept for environmental risks [29].

This chapter (and presumably most of this book) described chemicals as the source of "bad," but most chemicals are produced and used because they are "good" for life and well-being of the people. We need to realize that we accept the potential risk of using chemicals in exchange for the benefit of using it. Then, the core issue may not be the risk per se but the fact that we accept the risk without knowing it, which arises from lack of risk as well as benefit information, which in turn arises not only from lack of scientific data but also from insufficient dissemination of message.

As discussed in this chapter, environmental risk issues gradually become an area of complicated issues. *Systems approach*, which focuses on the relationship among *multiple systems* associated with the concerned problem, might be a useful tool to tackle such a complicated (*wicked*) issue (see [30]).

References

1. Steffen W, et al. The trajectory of the Anthropocene; the great acceleration. Anthropocene Rev. 2015a;2:81–98.
2. Kruzen PJ, Stoermer EF. The 'Anthropocene'. Global Change News Letter. 2000;41:17–8.
3. UNESCO. Global Chemicals Outlook II, p.10, Figure 1.9. 2019.
4. Ministry of Environment, Japan. Annual report on the environment, 1988 Japan. Chapter 3. Section 2. 1(2) (1988) (in Japanese).
5. Wang Z, Altenburger R, Backhaus T, Covaci A, Diamond ML, Grimalt JO, Lohmann R, Schäffer A, Scheringer M, Selin H, Soehl A, Suzuki N. We need a global science-policy body on chemicals and waste – Major gaps in current efforts limit policy responses. Science. 2021; https://doi.org/10.1126/science.abe9090.
6. USEPA. Risk assessment -risk assessment basics. https://www.epa.gov/risk/about-risk-assessment#whatisrisk. Accessed 2021 April 17.
7. European Environmental Agency Report No.1, 2013; https://www.eea.europa.eu/publications/late-lessons-2. Accessed 2021 March 13.
8. Rockstrom J, et al. A safe operating space for humanity. Nature. 2009;462:472–5.
9. Steffen W, et al. Planetary boundaries: guiding human development on a changing planet. Science. 2015b;347(6223):1259855. https://doi.org/10.1126/science1259855.
10. Diamond ML, et al. Exploring the planetary boundary for chemical pollution. Environ Int. 2015;78:8–15.
11. Villarrubia-Gomez P, Corenell SE, Fabres J. Marine plastic pollution as a planetary boundary threat- the drifting piece in the sustainability puzzle. Mar Policy. 2018;96:213–20.
12. Bjorn A, Hauschild MZ. Introducing carrying capacity-based normalization in LCA: framework and development of references at midpoint level. Int J Life Cycle Assessment. 2015; https://doi.org/10.1007/s11367-015-0899-2.
13. Chuwdhury RB, Moore GA, Weatherley AJ, Arora M. Key sustainability challenges for the global phosphorus resource, their implications of global food security, and for mitigation. J Cleaner Production. 2017;140:945–63.

14. Budnik LT, Casteleyn L. Mercury pollution in modern times and its socio-medical consequences. Sci Total Env. 2019;654:720–34.
15. Frumkin H, Haines A. Global environmental change and noncommunicable diseases risks. Annu Rev Public Health. 2019;40:261–82.
16. Future Earth. Ten new insights in climate science. 5. Climate change can have profoundly affect our mental health. 2020. https://10nics2020.futureearth.org/10-new-insights-in-climate-science/5-climate-change-can-profoundly-affect-our-mental-health/. Accessed 2021 March 18.
17. Battilani P, Toscano P, Van der Fels-Klerx HJ, Moretti A, Camardo Leggieri M, Bera C, Rotais A, Goumperis T, Robinson T. Aflatoxin B1 contamination in maize in Europe increases due to climate change. Sci Rep. 2016;6:24328. https://doi.org/10.1038/srep24328.
18. Berdalet E, Fleming LE, Gowen R, Davidson K, Hess P, Backer LC, Moore SK, Hoagland P, Enevoldsen H. Marine harmful algal bloom human health and wellbeing: challenges and opportunities in the 21st century. J Marine Biol Assoc UK. 2016;96:61–91.
19. Behrens P, Kiefte-de Jong JC, Bosker T, Rodrigues JFD, de Koning A, Tucker A. Evaluating the environmental impacts of dietary recommendations. Proc National Acad Sci USA. 2017;114:13412–7.
20. Wild CP. Complementing the genome with and 'exposome'. The outstanding challenge of environmental exposure measurement in molecular epidemiology. Cancer Epi Biomarkers Prev. 2005;14(8):1847–50.
21. Maitre L, et al. Human early life Exposome (HELIX) study: a European population -based exposure cohort. BMJ Open. 2018;8:e021311. https://doi.org/10.1136/bmjopen-2017-021311.
22. Tamayo-Uria I, et al. The early-life exposome: description and patterns in sex European countries. Env International. 2019;123:189–200.
23. Vrijheid M, et al. Earlh-life environmental exposure and childhood obserivy: an Exposome-side approach. Environ Health Perspect. 2020;128:67009.
24. Kahn LG, Philippat C, Nakayama SF, Slam AR, Trasande L. Endocrine-disrupting chemicals: implications for human health. Lancet Diabetes Endoc. 2020;8:703–18.
25. Watanabe C, et al. Males in rural Bangladeshi communities are more susceptible to chronic arsenic poisoning than females: analyses based on urinary arsenic. Environ Health Perspect. 2001;190:1265–70.
26. Escher BI, Hackermuller J, Polte T, Scholz S, Aigner A, Altenburger R, et al. From the exposome to mechanistic understanding of chemical-induced adverse effects. Environ Int. 2017;99:97–106.
27. Ciallella HL, Zhu H. Advancing computational toxicology in the big data era by artificial intelligence; Data-driven and mechanism-driven modeling for chemical toxicity. Chem Res Toxicol. 2019;32:536–47.
28. Muratov EN, et al. QSAR without borders. Chem Soc Rev. 2020;49:3525.
29. Whitmee S, et al. Safeguarding human health in the Anthropocene epoch; report of the Rockefeller foundation – Lancet commission on planetary health. The Lancet. 2015;386:1973–2028.
30. Dyball R, Newell B. Understanding human ecology. A systems approach to sustainability. Abington, UK: Routledge; 2015.

Chapter 19
Spectrum of Environmental Risk Assessment, Management, and Communication

Mari Asami

Abstract Environmental risks have a broad spectrum of issues. Effluent from mines, industrial wastewater, contaminated fish and rice, and polluted air have resulted in severe health problems worldwide, especially since the late 1950s. The Government of Japan has made tremendous efforts to establish and improve regulations to protect the environment. In many cases, effective countermeasures took a long time to implement as there was a need for concrete scientific evidence. It is necessary to continuously improve environmental risk assessment and management frameworks, not only to revise standards but also to improve the methods used for risk assessment, management, and communication, mainly based on scientific risk assessment/management schemes driven by precautionary approaches. The policies should include further consideration of the future environment and their social impacts. In addition, factors related to financial and management sustainability should be considered, as there is a severe decrease in population in many local areas in Japan. Careful mutual risk communication and consensus formation are important to optimize regulation.

Keywords Environmental risk · Regulation · Environmental disease burden Precautionary approach

M. Asami (✉)
Department of Environmental Health, National Institute of Public Health, Saitama, Japan
e-mail: asami.m.aa@niph.go.jp

T. Nakajima et al. (eds.), *Overcoming Environmental Risks to Achieve Sustainable Development Goals*, Current Topics in Environmental Health and Preventive Medicine,
https://doi.org/10.1007/978-981-16-6249-2_19

19.1 Introduction

The environment is the basis for public, community, and individual health. Human health is largely impacted by the environment. Many diseases and health disorders can be prevented by focusing on environmental risk factors. The World Health Organization (2016) [1] systematically analyzed and quantified how environmental risks affect a wide range of diseases and examined the regions and populations most vulnerable to environmentally mediated death, disease, and injury. They covered the impacts of environmental risks on health for more than 100 diseases and injuries, including unsafe drinking water and sanitation, air pollution, and indoor stoves. They highlighted areas where immediate intervention is desirable and deficiencies where more research is needed, established linkages, and quantified the burden of disease associated with a variety of environmental risk factors. The study on the impact of environmental risk factors on global disease and injury showed that environmental risks account for a large proportion of the global burden of disease. That is, the risk attributable to preventable environmental risks was estimated to be responsible for 23% (95% CI: 13–34%) of all deaths and 22% (95% CI: 13–32%) of the disease burden in terms of disabled life years (DALYs), a combined measure of life-years lost due to death and to disability (Fig. 19.1).

A number of immediate measures can be taken to reduce the burden of disease caused by environmental factors, such as promoting safe household water storage and better sanitation measures, the use of clean fuels, more judicious use and management of hazardous substances in homes and workplaces, and occupational safety

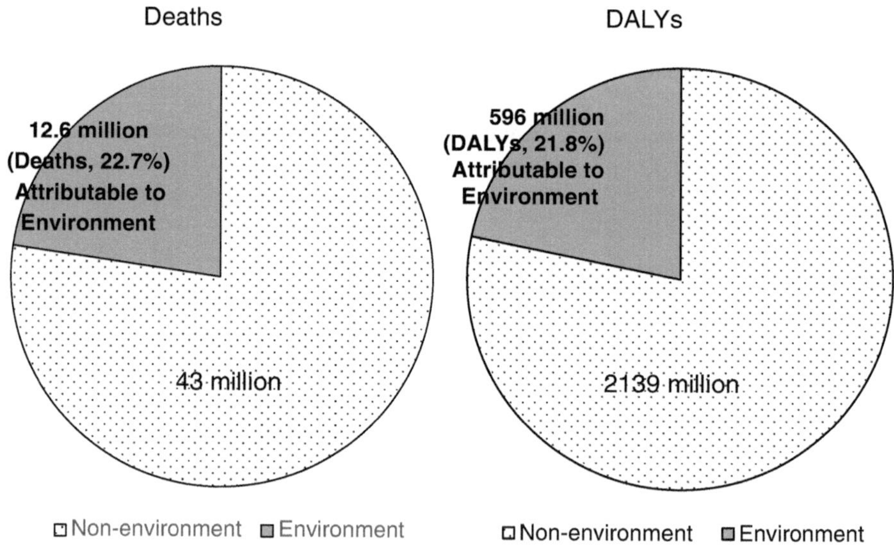

Fig. 19.1 Fraction of deaths and DALYs attributable to the environment globally in 2012

and health measures. It is also critical that sectors, such as energy, transport, agriculture, and industry, work with the health sector to address the root causes of environmental health hazards. Action needs to be taken through decision-making in all sectors that affect the environmental determinants of health, and not just in the health sector. A coordinated approach to health, environment, and development policy can enhance and sustain improvements in human well-being and quality of life through a number of social and economic co-benefits. Repositioning of the health sector and working across sectors on effective preventive health policies represent ways forward to address the environmental causes of disease and injury and could ultimately transform the global burden of disease.

The rate largely is affected by age. Figure 19.2 shows the fraction of environmental changes according to age. Children are especially affected by the environment, with high rates of environment-related injury [1]. DALYs of children under 5 years old are affected by environment causing injuries over 50%. Figure 19.3 compares the DALYs from preventable environmental risks, the proportion of disease attributable to the environment, and the main areas of environmental activity that can prevent disease. Large portion of infectious diseases, injuries, and even cardiovascular diseases can be prevented by the improvement of environment.

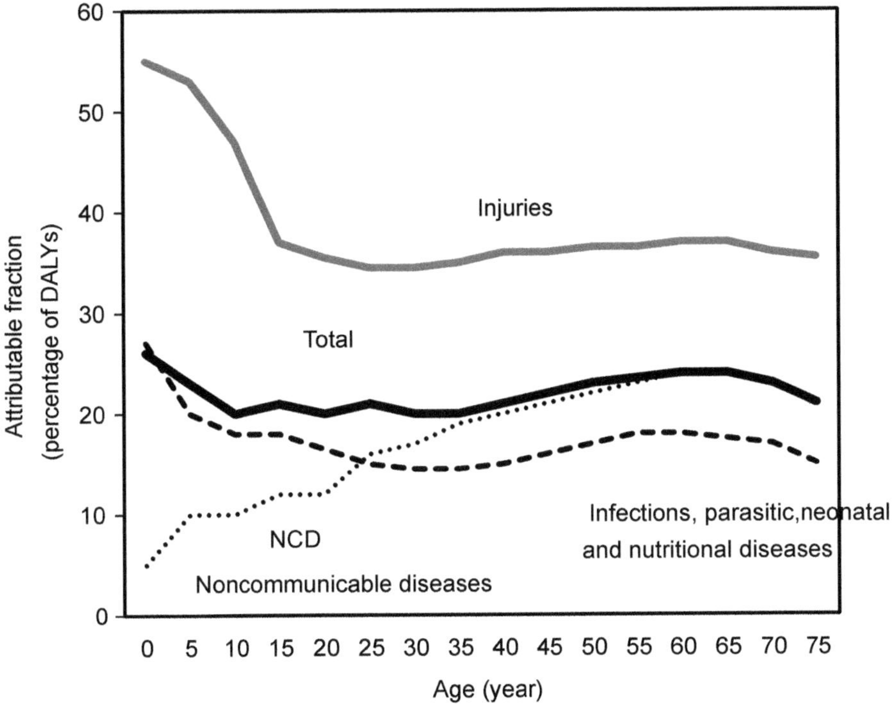

Fig. 19.2 Environmental fraction of global burden of disease (in DALYs) by age and disease group in 2012

Fig. 19.3 Diseases with the highest preventable disease burden from environmental risks in disability-adjusted life years (DALYs), a combined measure of years of life lost due to mortality and years of life lived with disability in 2012 [1]

19.2 Status of Health Conditions in Japan

The causes of death and burden of disease in Japan are summarized by Asami and Kunugita (2018) [2]. The estimated causes of the number of years of life lost and other statistics are collected. The numbers are based on the health statistics and also estimated by diseases. The top 25 causes of disease expressed as the number of years of life lost (YLL, in thousands) in 2010 in Japan are shown in Fig. 19.4 [3]. Stroke, ischemic heart disease, lower respiratory tract infections, lung cancer, and self-harm are the five top causes of YLLs, which are also affected by environmental conditions. Chronic obstructive pulmonary disease (COPD), road injuries, falls, and drowning are also strongly related to environmental factors. Injuries are usually not considered as causes of diseases, but they are major issues in children's health and triggers of increased health burden in the elderly.

Figures 19.4 and 19.5 show the causes of disease burden [3]. With regard to the causes of cardiovascular disease, diet ranked first, hypertension second, smoking third, physical inactivity fourth, and high body mass index (BMI) fifth, followed by high fasting plasma glucose, alcohol use, ambient particulate matter (PM) pollution, high total cholesterol, and occupational risks. Diet, tobacco, and alcohol drug use

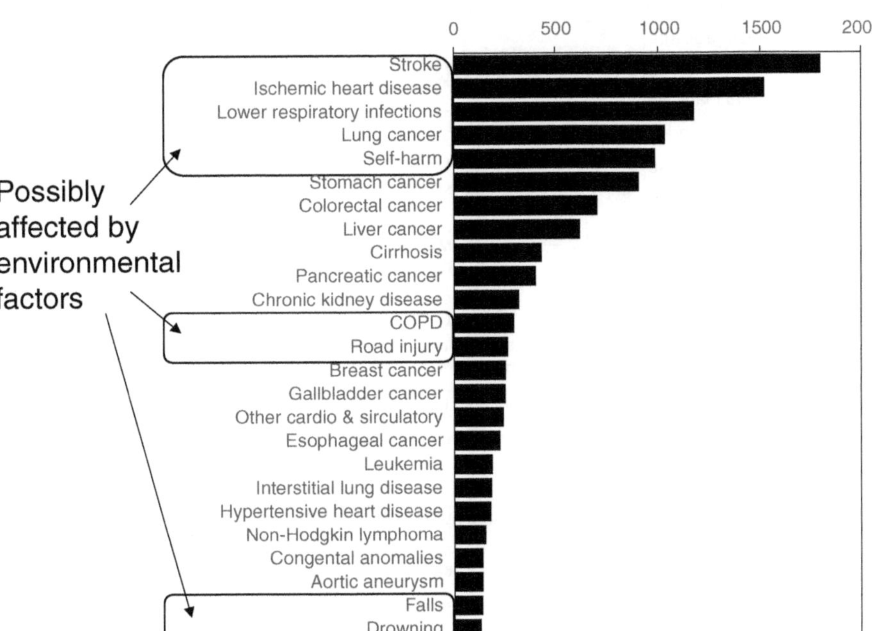

Fig. 19.4 Top 25 causes of YLLs in Japan in 2010 [3]

were listed as lifestyle factors, and labor risk and air pollution as environmental/labor risks. Thus, dietary risk and hypertension account for a large proportion of the causes of cardiovascular disease.

19.3 Previous Pollution Issues in Japan

The Committee on Environmental Risk, Science Council of Japan, made a report on "Regulatory science on environmental risks for decision making of environmental policies" [4]. In the report, the committee pointed out the previous pollution issues in Japan. Itai-itai disease related to chronic cadmium poisoning due to environmental pollution with cadmium that is characterized by proximal tubular osteochondrosis became a problem in the Jinzu River basin, Toyama Prefecture, in 1955. In 1920, the then Chairman of the Board of Agriculture submitted a proposal to the government and the Governor of Toyama Prefecture for removal of mineral poisoning, which suggests that long-term contamination was recognized at this time. Multiple proximal tubular dysfunctions (hereinafter referred to as cadmium nephropathy) occur frequently in residents of cadmium-contaminated areas in the Jinzu River basin, but even today the Ministry of the Environment does not authorize "cadmium

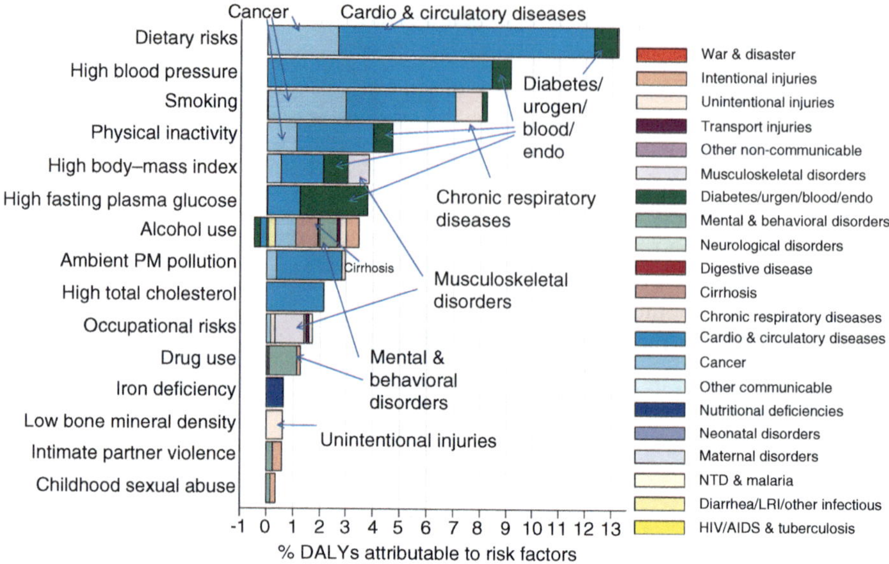

Fig. 19.5 Burden of disease attributable to 15 leading risk factors in 2010 expressed as a percentage of Japan DALYs [3]

nephropathy" itself as a pollution-related disease. Rice cultivation in the paddy fields along the lower reaches of the Jinzu River had already been damaged at the end of the Meiji era (1868–1912), and farmers had often protested and demanded countermeasures from the mining company and the government. Based on the results of research, the Ministry of Health and Welfare announced in May 1968 that "Itai-itai disease is a chronic poisoning caused by cadmium discharged by the Kamioka Mine and is a pollution disease."

The first report of Minamata disease was made in 1956, and in March 1957, the Health and Welfare Science Research Group reported that "the most suspected case is (omission) poisoning due to ingestion of fish and shellfish caught in Minamata Bay Harbor," but no identification on specific agents, or original sources was made then. Under the administrative guidance of Kumamoto Prefecture, the Minamata City Fishermen's Cooperative Association began self-regulating fishing in Minamata Bay in August 1957. On September 26, 1968, the Ministry of Health and Welfare and the Agency for Science, Technology and Research issued a unified government opinion, stating that Minamata disease in Kumamoto and Niigata was caused by methylmercury compounds produced as a byproduct of industrial manufacturing processes. In addition, air pollution and other industrial contamination were severe in the 1960s–1980s. The trend of registered and supported cases by the government as public nuisance is described in Fig. 19.6.

The adverse public health effects and economic damage continued, as measures were not put in place to stop the pollution until the cause and mechanism of action had been conclusively determined, even after the occurrence of the disease in patients. A report by the Ministry of the Environment [5] presented an example in

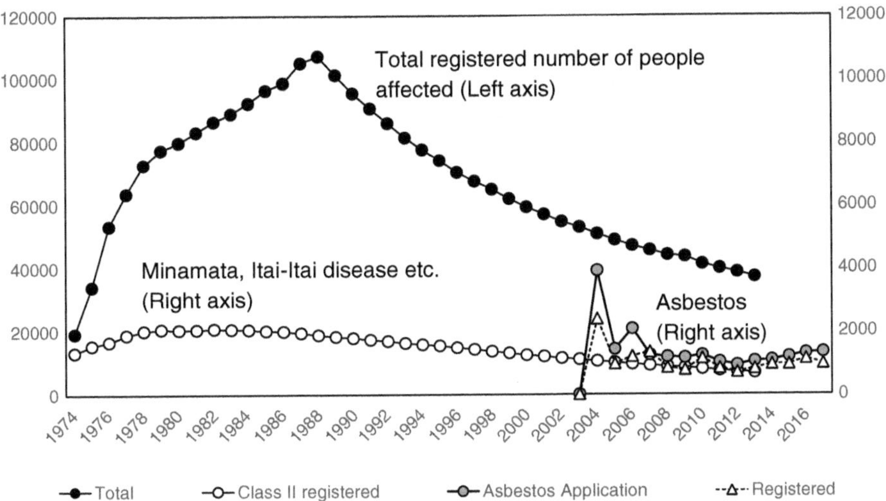

Fig. 19.6 Registered and supported people affected by environmental pollution, derived from [5]

which the delay in adopting countermeasures resulted in increased damage, and the resulting costs (annual compensation, dredging, and compensation for damage to fisheries) were estimated to be about 100 times (12.6 billion yen) the annual pollution prevention investment in the plant that caused the problem.

Relatively recently, in 2005, it was discovered that 51 people had died in the past 10 years due to asbestos-related diseases at a factory producing asbestos pipes. Subsequently, it was discovered that mesothelioma had occurred among factory workers and residents in the neighboring area. This became a major legal case, with compensation payments made to 300 people who were thought to have been causally affected [6].

19.4 Individual Issues in Japan

19.4.1 Issues Regarding Chemical Pollutants in Japan

Historically, countermeasures for pollution-related problems and compensation for those affected have only been implemented after adverse health effects have already occurred and been widely publicized. Here, we discuss the necessary frameworks to counteract the environmental risk of chemically induced occupational diseases due to unknown toxicity, focusing mainly on solvent intoxication.

Kamijima and Shibata (2018) [7] reported the importance of occupational illnesses in forecasting environmental risks. For example, in the 1950s, frequent occurrences of chronic benzene intoxication led to the restriction of benzene use as a solvent. Benzene was confirmed to be a carcinogen and was eliminated from gasoline. Benzene was replaced by *n*-hexane, which was considered safer at the time, but which in turn was shown to be responsible for frequent occurrences of peripheral polyneuropathy. Incorporating industrial health issues into environmental health risk analysis seems particularly important.

In March 2012, the Ministry of Health, Labour and Welfare (MHLW) received a claim from a printing factory in Osaka Prefecture that a person had developed cholangiocarcinoma due to the use of chemical substances, and in September 2012, the Study Group on Non-occupational Causes of Cholangiocarcinoma Occurring at a Printing Factory began examining the causal relationship with the working environment. In March 2013, it was shown that the cholangiocarcinoma that developed in the workers at this workplace was most likely caused by long-term exposure to high concentrations of 1,2-dichloropropane [7].

Trichloroethylene was used extensively as a solvent because it has low ozone-depleting potential to substitute other chemicals, as well as 2-bromopropane used in Korea. In 2012, a cluster of cases of cholangiocarcinoma among workers in an off-set proof-printing plant became a matter of public concern. It was concluded that the disease was due to long-term and high-level exposure to 1,2-dichloropropane or dichloromethane. The extensive exposure of the affected workers was attributed to the lack of regulation regarding enforced environmental measurements and health checkups for workers exposed to 1,2-dichloropropane [7]. The substance was not regulated at that time; however, appropriate information sharing and intensive speedy consideration should be implemented into the regulation.

The chemical 2-ethyl-1-hexanol is detected at high concentrations in a type of room with a certain interior finishing work and is likely a cause of sick building syndrome. It was found that 2-ethyl-1-hexanol, a highly volatile compound, is emitted from flooring due to hydrolysis of compounds, including phthalate plasticizers such as di-2-ethylhexyl phthalate (DEHP), in flooring materials made of plastics when they come into direct contact with concrete. It is also a matter of concern which is unintentional contamination from a by-product.

In addition, there are still many cases of health problems caused by asbestos and 1,2-dichloropropane, which could have been reduced if prompt investigation and countermeasures had been implemented. Figure 19.7 shows the scheme of regulation of chemicals related to occupational health and environmental management. We should pay careful attention to occupational health issues to tackle with insufficient exposure control unknown occupational toxic substances and their by-products, which potentially further affects human health through environment.

19.4.2 Importance of Early Exposure Assessment and Management

Early exposure assessment and implementation of countermeasures are important. Here, we describe an example of a rapid response to the recent outbreak case of illness caused by a chemical substance. The contamination of groundwater with organic arsenic in Kamisu, Ibaraki Prefecture, was discovered in 2003 [2]. As there were clear adverse health effects due to ingestion of contaminated water, the patients

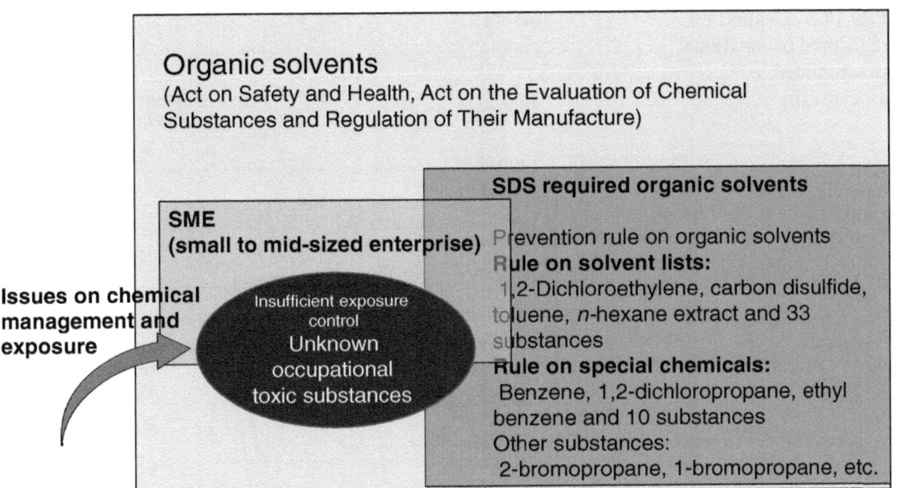

Fig. 19.7 Problems in chemical management in fields of medicine and engineering [7]

were identified quickly. The medical handbook that ensured official support of medical treatment was distributed based only on residency and exposure requirements after the discovery. It is very important to ensure the accumulation of such objective data and the transparency of the response. If it is determined based only on diagnosis and symptoms, it may take more time and sometimes require a lawsuit as in the case of Itai-itai disease cases and in Minamata in the 1950s.

Although means of chemical management enforced by laws have been developed, exposure management based on the assumption that the chemical in use has unknown toxicity is necessary, especially if the chemical was commercially available when the law was first enforced. Cooperation between specialists in the fields of medical science and technology is necessary to resolve the problem of exposure to unexpected chemicals. The authors propose the advancement of curriculums in schools of technology in Japan from the perspective of health, i.e., toxicity of chemicals. In addition, attention must be paid to the occurrence of chemically induced occupational diseases in industrializing countries that have still yet to need to develop substantial occupational safety and health measures.

19.4.3 Asbestos

Asbestos is still an emerging concern in Japan. Figure 19.8 shows registered occupational cases of mesothelioma, registered occupational cases of lung cancer, and the annual number of deaths by mesothelioma in Japan [6]. As mesothelioma is known to be mainly caused by asbestos exposure, the number of registered cases is limited to some extent compared to the number of deaths due to mesothelioma. Adverse effects of asbestos are known to emerge as much as 20 years after exposure, and the number of cases and deaths may still increase.

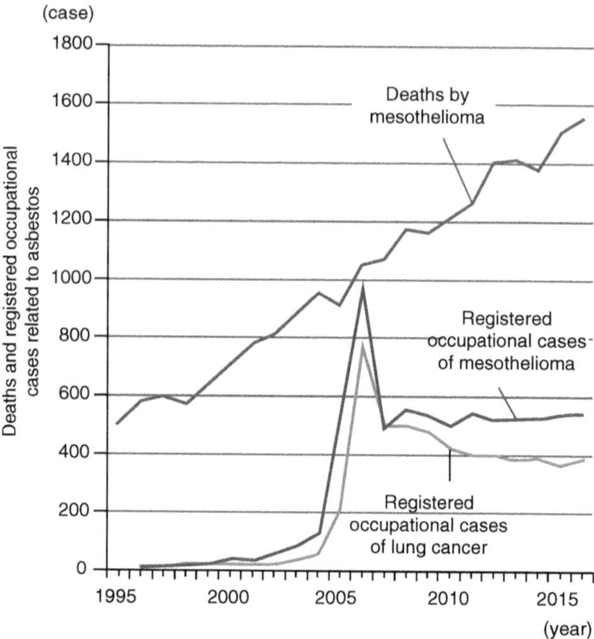

Fig. 19.8 Deaths and registered occupational cases related to asbestos [6]

19.4.4 Indoor Health Risks

Figure 19.9 shows the relations between environmental temperature and adverse health effects related to cardiovascular disease and respiratory disease as reported by Hayama et al. (2014) [8]. The monthly mortality rate is much higher when the monthly average temperature is low (<8 °C) compared to summer when the monthly average temperature is 8–25 °C. Therefore, it is recommended to heat bathrooms and other indoor areas in winter.

19.4.5 Multiple Environmental Risks

Figure 19.10 shows many of the direct and indirect causes of environmental health and ecological risks. We are surrounded by many risk factors through environment, not only by chemicals but also by physical, biological, or social factors largely affected human health. Direct causes are described in the captions of Fig. 19.10 in the list. Many environmental risks are pointed out and some of them are fatal, some of them are important and, some of them are ubiquitous and not directly health-related but inducing social concern. We must identify their risk and characteristics to prioritize environmental policy issues.

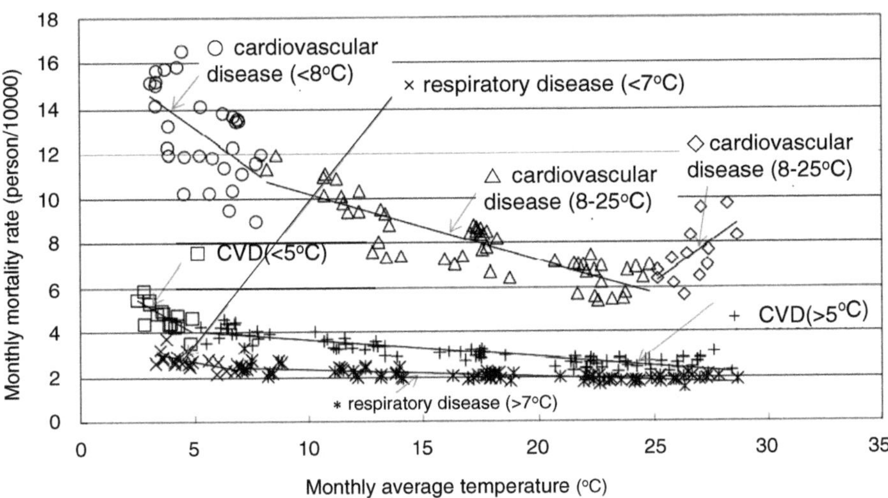

Fig. 19.9 Temperature and health [8]. *CVD* cerebral vascular disease

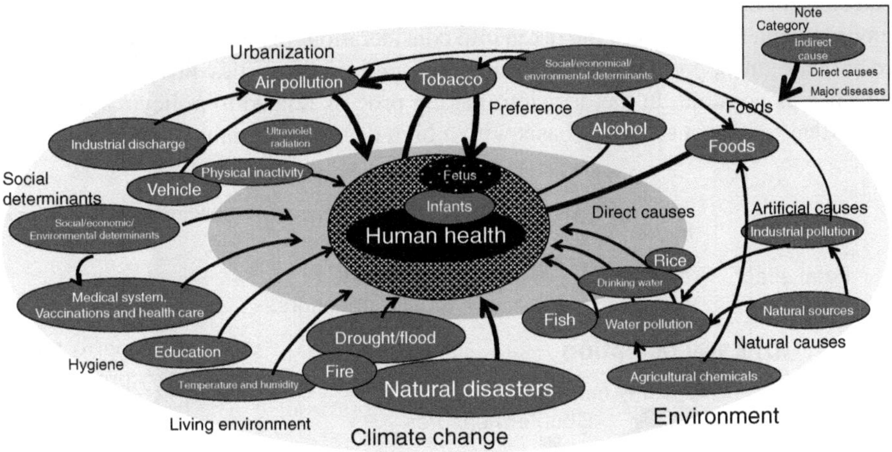

Fig. 19.10 Relations between main environmental risks and human health risks, modified from [2]

19.5 Risk Assessment and Precautionary Approach

The precautionary principle is often discussed in relation to environmental risks. The term "precautionary principle" is also used in treaties as a translation of the term meaning proactive measures which will be discussed elsewhere. Some have argued that the term should be replaced with "precautionary approach," "precautionary measures," "proactive measures," or "precautionary approach" [4]. In the

EU, the precautionary principle is stipulated in the Operational Treaty (EU Operational Treaty, Article 191, paragraph 2). Figure 19.11 shows hazard, risk and response, better controlled based on early detection, risk assessment, and smart forecast, based on the past, current, and ideal situations of risk assessment. The most important issue is fatal effect. In the past, it often took a long time to remedy cases of environmental pollution after the severe public health effects had become apparent using traditional countermeasures. Early detection would allow countermeasures to be taken earlier and would therefore decrease the adverse effects of environmental pollution. Appropriate risk assessment and application of the precautionary approach may decrease or completely abrogate the adverse effects of environmental pollution. To introduce the precautionary approach, we should be keen to find the sign of the adverse change of health or environment.

There are cases where mitigation may not be cost-effective or where such measures are unlikely to be effective due to the natural environmental conditions or technical constraints, e.g., in the fields of radiation protection and food safety. For example, in these fields, there are cases where the principle of As Low As Reasonably Achievable (ALARA) must be followed and measures must be taken by any means possible. In the field of engineering, Best Available Technology (BAT) may be selected. In addition, in cases where there is original contamination in the natural environment, this must also be taken into consideration. In such cases, it is desirable to provide an opportunity for coordination among the parties involved as far in advance as possible. In the decision-making process related to policymaking, it is important to have a scientific basis with a high degree of certainty and probability,

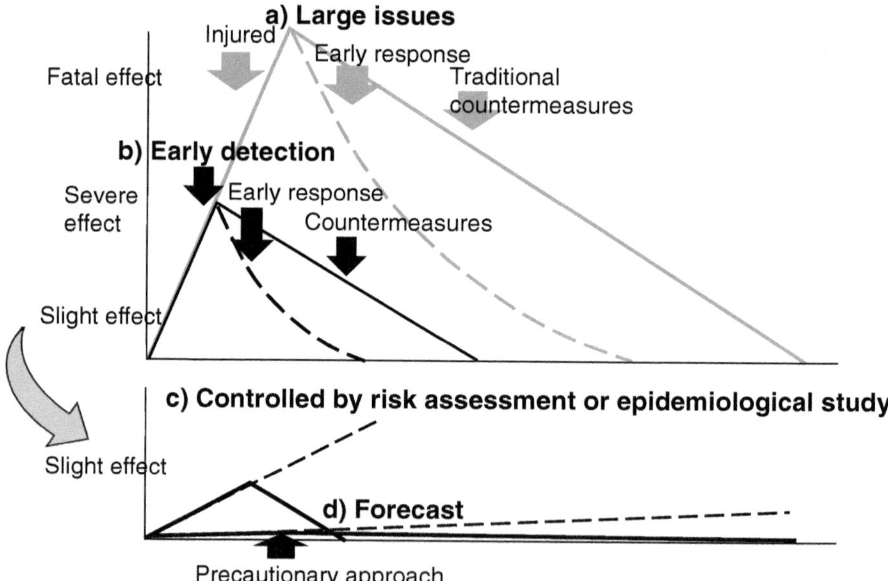

Fig. 19.11 Risk and response, better controlled based on early detection, risk assessment and smart forecast

as well as to compare and examine measures that can be achieved in a realistic and rational manner.

19.6 Precautionary Principle

19.6.1 Precautionary Principle Approach

The Rio Declaration on Environment and Development (the Rio Declaration) [9] was a short document produced at the 1992 United Nations Conference on Environment and Development (UNCED), informally known as the Earth Summit. The Rio Declaration outlined 27 principles intended to guide countries in future sustainable development. The articles include formulations of the precautionary principle, which should be "widely applied by states according to their capabilities" (principle 15), and of the "polluter pays principle."

The precautionary principle is outlined in the EU Operational Treaty (Article 191, paragraph 2). In Japan, although the Basic Environment Law does not explicitly state the precautionary principle, Article 4 of the same law is related to this principle. In addition, the Fourth National Strategy for the Environment stipulates the precautionary approach. The response to the Minamata disease incident in the 1950s and 1960s should have been really an issue of the precautionary principle.

In addition, the precautionary principle is applicable in dealing with global climate change, endocrine-disrupting chemicals, neonicotinoid pesticides, and microplastics in the environment. An EU Communication Paper published in 2000 [10], which presented the basic concept of the precautionary principle, stated that if a certain level of trust and reputation is recognized sufficient consideration should be given to risk assessments, even if they are in the minority in the scientific community. It has sometimes been stated that the precautionary principle and risk theory are opposed to each other, with the former taking the perspective of the affected party and the latter taking the perspective of the development side. However, risk theory does not necessarily presuppose development, and we should aim toward the integration of risk theory and the precautionary principle. According to the Ministry of the Environment (2013) [5], the cost of the damage caused by Minamata disease in the area around Minamata Bay was shown to be much higher than the cost of pollution control and preventive measures (12,631 vs. 123 million Japanese yen per year, respectively).

19.6.2 Reasonably Achievable Principle

On the other hand, there are cases where the cost of mitigation is extremely high from the viewpoint of cost-effectiveness, or where the effectiveness of practical measures is difficult in the fields of radiation protection and food safety. For example, the As Low As Reasonably Achievable (ALARA) principle has been introduced in the fields of radiation protection and food safety [2]. In engineering, Best Available Technology (BAT) is sometimes selected. In addition, endogenous contamination in the natural environment must also be taken into consideration. In such cases, it is desirable to provide an opportunity for coordination among the parties involved as far in advance as possible. In the decision-making process related to policymaking, it is important to have a scientific basis with a high degree of certainty and probability, as well as to compare and examine measures that can be achieved in a realistic and rational manner.

19.6.3 Environmental Responsibility

It has long been the case that the victim must prove intent or negligence on the part of the perpetrator in the event of environmental pollution. However, as it is extremely difficult for residents who have suffered health damage to clarify the negligence of the offending company, the introduction of the concept of no-fault liability has become necessary from the perspective of maintaining social justice.

Even if there is a possibility of a causal relationship with the health effects in patients suffering health damage, it is not appropriate for the patient to be denied compensation due to strict criteria, such as not being able to obtain testimony from colleagues, or if the criteria are made stricter in consideration of the compensation capacity (budget). In addition to the pollutant payment principle (PPP), the Precautionary polluter-pays principle (PPPP) has been proposed, which applies the precautionary principle, requires potential hazardous products to be taxed in advance, and the taxes are refunded if the products are proven to be safe.

19.6.4 Environmental Impact Assessment, Strategic Environmental Assessment, and the Use of Various Assessments

In addition to risk assessment, health and environmental policies are based on environmental impact assessment, policy assessment, regulatory impact analysis, and health impact assessment. Environmental impact assessment is a method that attempts to evaluate the impact of large-scale public projects on the environment, such as in environmental assessment.

The so-called strategic environmental assessment involves the assessment of policies, legislation, programs, and plans for projects and regional development, mainly related to land development, and has already been legislated in EU member countries, Canada, and the USA. Strategic environmental assessment is considered to have a complementary function to project assessment and to contribute to the realization of the principle of sustainable development. It is also useful for maximizing benefits and effects by balancing risks and costs. Of course, the effects and benefits include socioeconomic factors, and so they cannot be judged solely based on evidence from natural science. In this sense, a humanistic and social science approach that goes beyond simple "scientific effects" is necessary. Another feature of this assessment is that it provides sufficient cumulative environmental impact assessment, which is inadequate in project assessment for a single project.

In Japan, environmental assessment in public projects was introduced in 1972, and later the promotion of environmental assessment was positioned in the Basic Environmental Law enacted in 1993 and the Environmental Impact Assessment Law was enacted in June 1997. Environmental assessment is effective in assessing the impact of a project, but in effect it mainly assessed the physical environment to the living environment and the natural environment, focusing on the protection of biological species.

The 2010 amendment of the Environmental Impact Assessment Law introduced the Planning Stage Consideration Procedure. However, in Japan, the introduction of strategic environmental assessment is necessary and should be urgently considered to provide a transparent and objective environmental consideration process at the planning stage and to implement Article 19 of the Basic Environment Law.

With regard to other policy evaluations, standard guidelines on policy evaluation were established in 2001, and now policies related to large budgets require evaluation by each government ministry. The Guidelines for Conducting Prior Evaluation of Regulations was established in 2007 [11], and under the Policy Evaluation Act, for large-scale projects, regulatory impact analysis is required and is published on the ministries' websites. However, quantitative evaluation is often difficult as regulatory impact analysis is not always simple and many factors must be taken into account.

In addition, a method for predicting and assessing changes in health effects and factors related to health events (social determinants of health) that may be caused by proposed measures, projects, etc. has been proposed through health impact assessment as a means of emphasizing fairness in society and correcting social disparities, including health disparities. The Japanese Society of Public Health issued guidelines for such assessments in 2011 [12]. However, these assessment methods require a great deal of discretion, and they have been introduced at a practical level in only a limited number of cases to date in Japan.

The Guidelines for Environmental and Social Considerations were published in 2010 to promote the implementation of appropriate environmental and social considerations, especially human health and safety and inartificial conditions, particularly through the environment [13]. In addition to normal health issues, population movement, including involuntary resettlement, changes of land use, and local resource use may affect human health, especially socially vulnerable groups, including the poor and indigenous peoples. The report mentions that the assessment should focus on what is necessary for each individual project, including the distribution of damage and benefits, equity in the development process, gender, children's rights, cultural heritage, local conflicts of interest, infectious diseases, such as HIV/

AIDS, and the working environment. This is a more in-depth assessment of social aspects than general environmental impact assessments in Japan. This is in line with the Sustainable Development Goals (SDGs) of the United Nations and must be an important perspective in the evaluation of future environmental policies.

Risk is a concept expressed as the magnitude of a hazard multiplied by its probability of occurrence (i.e., its expectation value). The most widely used definition of risk communication is that used in a 1989 document published by the National Research Council (1989) [14], which refers to "the interactive process of exchanging information and opinions among individuals, institutions, and groups." "Information and opinions" in this definition includes "messages about the nature of risks" and "interest, opinions, and reactions to the development of laws and institutions for risk management." This does not imply mere disclosure of information. Disclosure of information by government agencies, academic councils, companies, etc., or press releases is unidirectional, but risk communication requires a response to the nature of the risk, the development of laws and systems, and the interest, opinions, and reactions of those who are informed. Kinoshita stated that the purpose of risk communication is "not one-way propaganda, but a two-way exchange of information and opinions among stakeholders (i.e., interested parties), and through this, to create a foundation on which the parties concerned can think together."

The role of mass media has been increasing, along with the spread of various social networking services, or social media, and their increasing influence. There have been a number of cases of health hazards triggered by newspaper reports, such as the problem of multiple cases of bile duct cancer among printers. The dioxin problem that began in 1998 was triggered by television reports. The Great East Japan Earthquake of 2011 and the resulting nuclear power plant accident, as well as the responses of Japan and the rest of the world, have been greatly influenced not only by existing mass media but also by social media using the Internet. In the conventional framework of environmental risk management, the roles of mass media and social media in risk communication have not been clearly defined. To date, risk communication has dealt with a single environmental risk at a time (e.g., toxic chemicals, etc.), and discussions have been conducted on the premise that stakeholders directly exchange information and opinions. The media is literally a "medium for intermediation," but the risk management framework did not envision cases where information is exchanged through the mass media. However, as the influence of the media has grown and how it communicates has become directly influential in policy formation, it is necessary to define its roles in some way.

There is another reason for the need for a new framework: there has been a major shift in sustainability research, including environmental risks. Future Earth, a worldwide framework created by the reorganization of international programs in global environmental research calls for trans-disciplinary efforts that transcend the boundaries between academia and society in promoting research activities to realize global sustainability. This means that not only academic experts but also various stakeholders in society will participate in the activities of Future Earth to harmonize science and technology with human beings and ecosystem in environment. The foundations of this effort are the exchanges of information and opinions among stakeholders. In risk communication for this purpose, it is essential for various stakeholders to question the various risks affecting the global environment and how science and technology can be used to solve them.

Risk assessment and risk management schemes have been introduced in a number of regulatory efforts. For example, the framework of food safety was established, as depicted in Fig. 19.12 [15]. Independence of risk assessment is ensured by the law, i.e., Food Safety Commission performs independent risk assessment based on scientific evidence, and ministries and other organizations introduce its assessment into risk management. We should establish the schematic framework in this manner, not only in food safety but also in environmental risk assessment.

19.7 Importance of Risk Communication

In case of emergency, crisis communication is necessary to persuade people to move to refuses or not to drink water and so on. It is rather different from normal risk communication. Health and ecological risks become obvious where natural hazards, vulnerability, and exposure overlap. Figure 19.13 shows the concept of the response to the hazard on chemical exposure. In the emergency case, countermeasures to control human exposure should be strongly recommended; however, if it is not so severe, care communication is needed.

Figure 19.14 is the trend of risk analysis and risk communication (an original image by the author). If the damage is severe and if the peak is so large, the

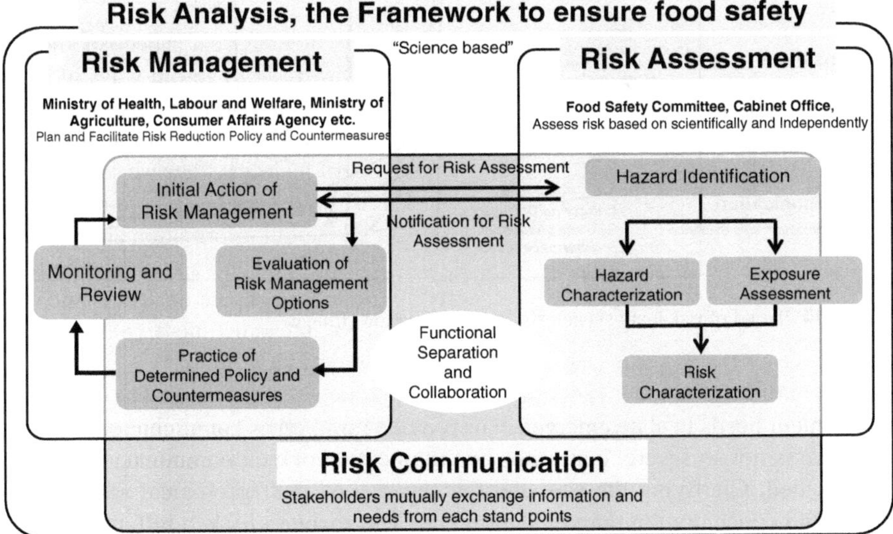

Fig. 19.12 Science-based risk analysis in food safety [15]. Sato (2020) arranged from Food Safety Analysis (WHO/FAO 2006)

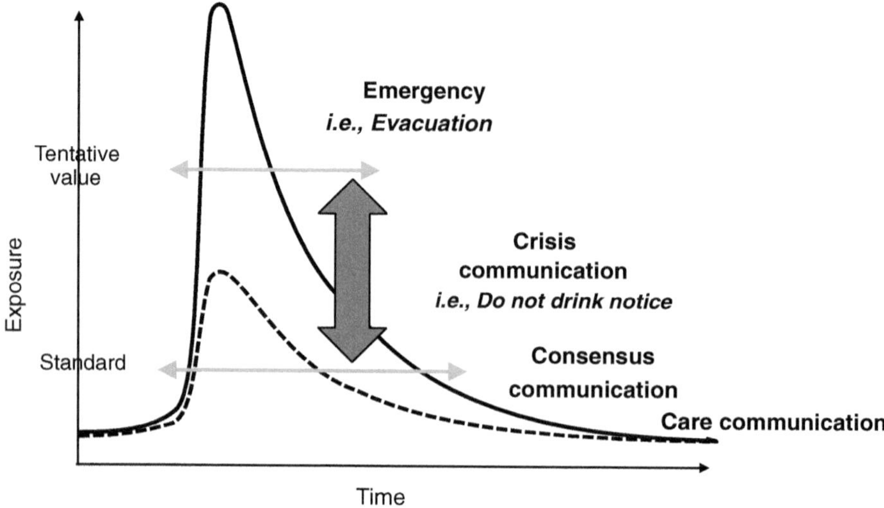

Fig. 19.13 Response to chemical accidents and risk communication

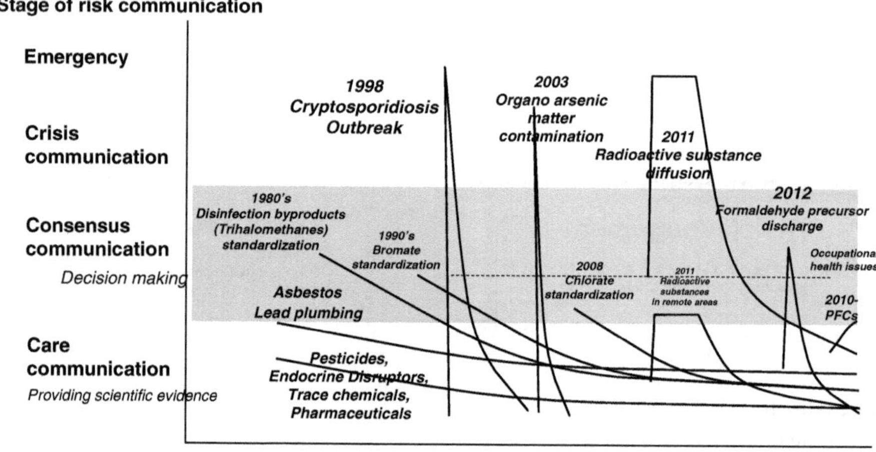

Fig. 19.14 Trend of risk analysis and risk communication (Image)

government needs to state emergency or provide rapid crisis communication. If the exposure is not so severe, consensus communication or care communication must be provided. Clarifying the size and the severity of risk (geological range, time range, and concentration range) is important. The shape (a peak, a hill, or a mountain), which is the trend of the issue, depends on the matter, and we must identify and foresee its trend.

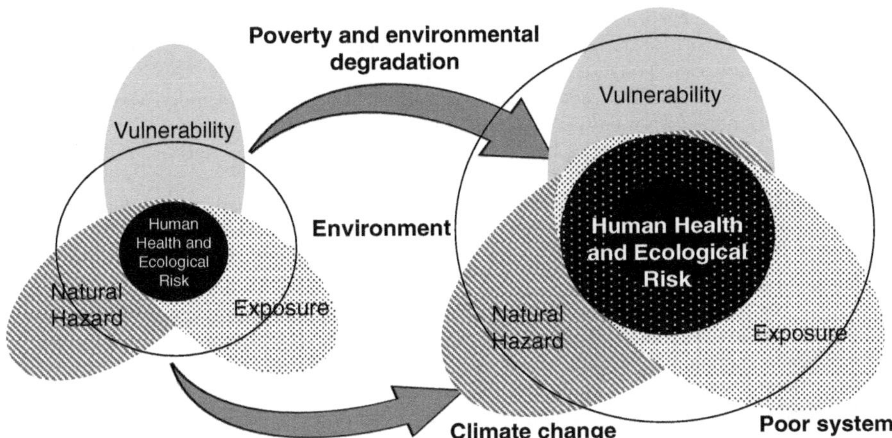

Fig. 19.15 Larger human health and ecological risk

19.8 Conclusion

Based on tragic experience and accumulation of scientific knowledge, we need to collect data, read literatures and foresee its risks. Appropriate management and communication must be undertaken to avoid severe risk (through crisis communication), mitigate and/or control risk (through consensus communication) or accept risks (through care communication). Health and ecological risks become obvious where natural hazards, vulnerability, and exposure overlap. The problems are becoming more severe because natural hazards, vulnerability, and exposures are expanding. Therefore, human health and ecological risks are increasing as outlined in Fig. 19.15 [16]. It is necessary to address future risks based on the precautionary approach.

Acknowledgments This research was supported by the Environment Research and Technology Development Fund (JPMEERF18S11705) of the Environmental Restoration and Conservation Agency of Japan.

References

1. World Health Organization. Preventing disease through healthy environments: a global assessment of the burden of disease from environmental risks. Geneva: WHO; 2016.
2. Asami M, Kunugita N. Various health risks attributable to environment. J Natl Inst Public Health. 2018;67(3):241–54.
3. Institute for Health Metrics and Evaluation. Global burden of diseases, injuries and risk factors study 2010. http://www.healthdata.org/sites/default/files/files/country_profiles/GBD/ihme_gbd_country_report_japan.pdf.

4. Science Council Japan, Committee on Environmental Risk. Report "Regulatory science on environmental risks for decision making of environmental policies", 2017 November. http://www.scj.go.jp/ja/info/kohyo/division-16.html
5. Ministry of Environment. Lessons from Minamata Disease and Mercury Management in Japan, 2013. https://www.env.go.jp/chemi/tmms/pr-m/mat01/en_full.pdf
6. Terazono A. Environmental risk by asbestos and future challenges. J Natl Inst Public Health. 2018;67(3):268–81.
7. Kamijima M, Shibata E. Countermeasures against unforeseeable chemically induced occupational illnesses as future environmental risks. J National Inst Public Health. 2018;67(3):282–91.
8. Hayama H, Saito M, Mikami H. Thermal environment for health and safety. J Natl Inst Public Health. 2014;63(4):383–93.
9. United Nations. Report of the United Nations conference on environment and development (Rio de Janeiro, 3–14 June 1992). https://www.un.org/en/development/desa/population/migration/generalassembly/docs/globalcompact/A_CONF.151_26_Vol.I_Declaration.pdf
10. EU. Commission adopts Communication on Precautionary Principle. EU Communication Paper IP/00/96, 2000. https://ec.europa.eu/commission/presscorner/api/files/document/print/en/ip_00_96/IP_00_96_EN.pdf
11. Ministry of General Affairs. Guideline for preassessment of regulation. 2007. https://www.soumu.go.jp/main_content/000499513.pdf
12. Japanese Society of Public Health. Guideline of Japan Association for Public Health. 2011.
13. Japan International Cooperation Agency. Guidelines for environmental and social considerations. 2010. https://www.jica.go.jp/environment/guideline/
14. National Research Council. document published by the National Research Council. 1989.
15. Sato H. Exposure Assessment in Risk Assessment. Gakujutsu no Doko (*in Japanese*), 2020;11:74–79. Originally notified at the glossary of Food Safety Commission, Japan, arranged from Food Safety Analysis (WHO/FAO, 2006). English translation was made by the author of this article (M Asami) referring the English Broucher of FSC, at https://www.fsc.go.jp/english/index.data/Broucher2018.pdf
16. World Bank. (Modified from world bank report) Climate Change and Natural Disasters in small Island Developing States. https://www.worldbank.org/content/dam/Worldbank/Climate%20change%20and%20natural%20disasters.pdf

GPSR Compliance

The European Union's (EU) General Product Safety Regulation (GPSR) is a set of rules that requires consumer products to be safe and our obligations to ensure this.

If you have any concerns about our products, you can contact us on ProductSafety@springernature.com

In case Publisher is established outside the EU, the EU authorized representative is:

Springer Nature Customer Service Center GmbH
Europaplatz 3
69115 Heidelberg, Germany

Batch number: 10370708

Printed by Printforce, the Netherlands